T0252138

Breakthroughs in Geology

Other earth science titles by Graham Park published by Dunedin include:

Mountains: The origins of the Earth's mountain systems (2017)
ISBN: 9781780460666

The Making of Europe: A geological history (2014)
ISBN: 9781780460239

Introducing Tectonics, Rock Structures and Mountain Belts Second Edition (2020)
ISBN: 9781780460949

Introducing Natural Resources (2015)
ISBN: 9781780460482

Introducing Geology: A Guide to the World of Rocks. Third Edition (2018)
ISBN: 9781780460758

For further details of these and other
Dunedin Earth and Environmental Sciences titles see
www.dunedinacademicpress.co.uk

Breakthroughs in Geology
Ideas that transformed earth science

Graham Park

EDINBURGH ◆ LONDON

Published in the United Kingdom by
Dunedin Academic Press Ltd
Head Office:
Hudson House, 8 Albany Street, Edinburgh, EH1 3QB
London Office:
352 Cromwell Tower, Barbican, London, EC2Y 8NB

www.dunedinacademicpress.co.uk

ISBNs
9781780460765 (Hardback)
978-1-78046-624-8 (PDF)
9781780466149 (ePub)
9781780466156 (Kindle)

© Graham Park 2020

The right of Graham Park to be identified as the author of this Work has been asserted by him in accordance with sections 77 and 78 of the Copyright, Designs and Patents Act 1988

All rights reserved.
No part of this publication may be reproduced or transmitted in any form or by any means or stored in any retrieval system of any nature without prior written permission, except for fair dealing under the Copyright, Designs and Patents Act 1988 or in accordance with the terms of a licence issued by the Copyright Licensing Society in respect of photocopying or reprographic reproduction. Full acknowledgment as to author, publisher and source must be given. Application for permission for any other use of copyright material should be made in writing to the publisher.

British Library Cataloguing in Publication data
A catalogue record for this book is available from the British Library

Typeset by Makar Publishing Production, Edinburgh
Printed in Poland by Hussar Books

Dedication

To my good friend and colleague, the late Professor John Tarney,
with whom I have had so many rewarding discussions over a
period of five decades, and who sadly passed away during the
preparation of this book.

Contents

Sourced illustrations

The following illustrations are reproduced by permission

Shutterstock: figures 1.1A, 1.1B, 1.2, 1.3, 1.4,2.1, 2.2A, 2.2B, 2.3, 2.5, 2.7, 3.8, 7.1A, 7.1B, 7.1C, 7.5.

Figure 1.4, 12.2 IPR/73-34C British Geological Survey ©NERC. All rights reserved.

Figure 2.6 Unknown, Hornet Magazine, via University College, London, digital (Public Domain).

Figure 3.1 Miyashiro, A., Aki, K. & Şengör, A.M.C. (1979) Orogeny, Chichester, John Wiley; Figure 1.1.

Figure 3.2, 3.9 Wegener, A. (1924) *The Origin of Continents and Oceans.* London, Methuen & Co.

Figure 3.11. Du Toit, A. (1937) *Our Wandering Continents. An Hypothesis of Continental Drifting.* London, Oliver & Boyd.

Figure 5.1 © Jim Wark.

Figure 6.8 Wilson, J.T. (1965) A new class of faults and their bearing on continental drift. *Nature,* London 207, 343–347.

Figure 8.6 Bally, A.W., Gordy, P.L. and Stewart, G.A. (1966) Structure, seismic data and orogenic evolution of the southern Canadian Rockies. *Canadian Association of Petroleum Geologists, Bulletin 14, 337–381.*

Figure 11.6 Mitchum, R.M., Jr., Vail, P.R. and Thompson, S., III (1977) Seismic stratigraphy and global changes of sea level, Part 2: the depositional sequence as a basic unit for stratigraphic analysis. In Payton, C.E. (ed.) *Seismic stratigraphy – applications to hydrocarbon exploration.* American Association of Petroleum Geologists, Memoir 26, Tulsa, Oklahoma, 53–62.

Figure 11.7 Mitchum, R.M., Jr., Vail, P.R. and Sangree, J.B. (1977) Seismic stratigraphy and global changes of sea level, Part 6: stratigraphic interpretation of seismic reflection patterns. In Payton, C.E. (ed.) *Seismic stratigraphy – applications to hydrocarbon exploration.* American Association of Petroleum Geologists, Memoir 26, Tulsa, Oklahoma, 117–133.

Figure 11.12 Vail, P.R., Mitchum, R.M., Jr., and Thompson, S., III (1977) Seismic stratigraphy and global changes of sea level, Part 4: global cycles of relative changes of sea level. In Payton, C.E. (ed.) *Seismic stratigraphy – applications to hydrocarbon exploration.* American Association of Petroleum Geologists, Memoir 26, Tulsa, Oklahoma, 81–97.

Figure 12.4 Ramberg, H. (1963) Experimental study of gravity tectonics by means of centrifuged models. *Geological Institutions of the University of Uppsala, Bulletin 42.*

Figure 12.6 Ramberg, H. (1967) *Gravity, deformation and the Earth's crust.* Academic Press, London.

Sourced illustrations

The following illustrations have been adapted from published sources

Figure 2.4 Darwin, Charles (1859) *On the Origin of Species by Means of Natural Selection, or the Preservation of Favoured Races in the Struggle for Life.* London, John Murray.

Figure 3.1 Şengör, A.M.C. (1982) Classical theories of orogenesis. In Miyashiro, A., Aki, K. & Şengör, A.M.C. *Orogeny*, Chichester, John Wiley.

Figure 3.3 Du Toit, A.L. (1937) *Our Wandering Continents. An Hypothesis of Continental Drifting.* London, Oliver & Boyd.

Figure 3.4 Zeuner, F.E. (1958) *Dating the past* (4th edn) Methuen, London.

Figure 3.5, 3.6 Du Toit, A.L. (1937) *Our Wandering Continents. An Hypothesis of Continental Drifting.* London, Oliver & Boyd.

Figure 3.7B, 3.10 Wegener, A. (1924) *The Origin of Continents and Oceans.* London, Methuen & Co.

Figure 3.12 Runcorn, S.K. (1962) Palaeomagnetic evidence for continental drift and its geophysical cause. In: Runcorn, S.K. (ed.) *Continental drift.* New York, London, Academic Press.

Figure 3.13 McElhinny, N.W. (1973) *Palaeomagnetism and plate tectonics,* Cambridge University Press.

Figure 4.3, 4.4 Holmes, A. (1929) Radioactivity and earth movements. *Transactions of the Geological Society, Glasgow* 18, 559–606.

Figure 4.5, 4.6 Vening Meinesz, F.A. (1962) Thermal convection in the Earth's mantle. *In*: S.K. Runcorn (ed.) *Continental drift.* London, Academic Press.

Figure 5.4, 5.5 Flinn, D. (1962) On folding during three-dimensional progressive deformation. *Quarterly Journal of the Geological Society, London,* 114, 385–433.

Figure 5.6 Watterson, J. (1968) Homogeneous deformation of the gneisses of Vesterland, Southwest Greenland. *Meddeleser om Grønland* 175, 6, 72pp.

Figure 5.11 Ramsay, J.G. (1980) Shear zone geometry: a review. *Journal of Structural Geology* 2, 83–99.

Figure 6.1 Wyllie, P.J. (1976) *The way the Earth works,* Wiley, New York.

Figure 6.2 Wilson, J.T. (1963) Hypothesis of Earth's behaviour. *Nature* 198, 925–929.

Figure 6.3 Hess, H.H. (1962) History of ocean basins. In Engel, A.E.J. *et al.* (eds) (1962): *Petrologic studies: a volume in honor of A.F. Buddington.* Geological Society of America Boulder, Colorado.

Figure 6.4 Larson, R.L. & Pitman, W.C. (1972) *Bulletin of the Geological Society of America,* 83, 3645–3661.

Figure 6.5 Chadwick, P. (1962) Mountain-building hypotheses. In: S.K. Runcorn (ed.) *Continental drift,* Academic Press, New York; London.

Figure 6.6 Benioff, H. (1949) Seismic evidence for the fault origin of oceanic deeps. *Geological Society of America Bulletin* 60, 1837–1866.

Figure 6.7 Wilson, J.T. (1965) A new class of faults and their bearing on continental drift. *Nature, London* 207, 343–347.

Figure 6.9 Larson, R.L. & Pitman, W.C. (1972) *Bulletin of the Geological Society of America* 83, 3645–3661.

Figure 6.10 Isacks, B., Oliver, J. and Sykes, L.R. (1968) Seismology and the new global tectonics. *Journal of Geophysical Research* 73, 5855–5899.

Figure 6.11 McKenzie, D.P. & Parker, R.L (1967) The North Pacific: an example of tectonics on a sphere. *Nature, London,* 216, 1276–1279.

Figure 6.12 Dewey, J. (1972) Plate tectonics. In: *Continents adrift and continents aground: readings* *from Scientific American*, Freeman, San Francisco.

Figure 6.13 Vine, F.J. & Hess, H.H. (1970) In: *The Sea*, v.4, Wiley, New York.

Figure 6.14 Isacks, B., Oliver, J. and Sykes, L.R. (1968) Seismology and the new global tectonics. *Journal of Geophysical Research* 73, 5855–5899.

Figure 6.15 Girdler, R.W. & Darracott, B.W. (1972) African poles of rotation. *Comments on the Earth Sciences: Geophysics* **2**, (5), 7–15

Figure 7.2 Ewing, M. (1965) The sediments of the Argentine basin. *Quarterly Journal, Royal Astronomical Society* 6, 10–27.

Figure 7.3 Bott, M.H.P. (1971) *The interior of the Earth*. London, Edward Arnold.

Figure 7.4 Gass, I.G. (1980) The Troodos massif: its role in the unravelling of the ophiolite problem and its significance in the understanding of constructive plate margin processes. In Panayiotou, A. (ed.) *Ophiolites, Proceedings of the International Ophiolite Symposium, Cyprus, 1979*. Nicosia, Cyprus Geological Survey Department, 23–35.

Figure 7.6 Gass, I.G. (1968) Is the Troodos massif of Cyprus a fragment of Mesozoic ocean crust? *Nature* 220, 39–42.

Figure 7.7 Moores, E.M. and Vine, F.J. (1971) The Troodos massif, Cyprus and other ophiolites as oceanic crust: evaluation and implications. *Royal Society, London, Philosophical Transactions* 268A, 443–466.

Figure 7.8 Kusznir, N.J. and Bott, M.H.P. (1976) A thermal study of the formation of oceanic crust. *Geophysical Journal, Royal Astronomical Society* 47, 83–95.

Figure 7.10 Glennie, K.W., Boeuf, M.G.A., Hughes-Clarke, M.W., Moody-Stuart, M., Pilaar, W.F.H. and Reinhardt, B.M. (1974) Geology of the Oman Mountains. *Koninklijk Nederlands geologisch mijnbouwkundig Genootschap.* (3 vols.)

Figure 8.1 Anderson, E.M. (1942) *The dynamics of faulting*. Oliver and Boyd, Edinburgh.

Figure 8.2 Gwinn, V.E. (1964) Thin-skinned tectonics in the Plateau and Northwestern Valley and Ridge Provinces of the Central Appalachians. *Geological Society of America, Bulletin* 75, 863–900.

Figure 8.3 King, P.B. (1951) *Tectonics of Middle North America*. Princeton University Press, Princeton, New Jersey.

Figure 8.4 Gwinn, V.E. (1964) Thin-skinned tectonics in the Plateau and Northwestern Valley and Ridge Provinces of the Central Appalachians. *Geological Society of America, Bulletin* 75, 863–900.

Figure 8.5, 8.7, 8.8 Bally, A.W., Gordy, P.L. and Stewart, G.A. (1966) Structure, seismic data and orogenic evolution of the southern Canadian Rockies. *Canadian Association of Petroleum Geologists, Bulletin* 14, 337–381.

Figure 8.9, 8.10 A, B Dahlstrom, C.D.A. (1969) Balanced cross sections. *Canadian Journal of Earth Sciences* 6, 743–757.

Figure 8.11 Elliott, D. and Johnson, M.R.W. (1980) Structural evolution in the northern part of the Moine Thrust Zone. *Royal Society of Edinburgh, Transactions (Earth Sciences)* 71, 69–96.

Figure 8.12, 8.13, 8.14 Boyer, S.E. and Elliott, D. (1982) Thrust systems. *American Association of Petroleum Geologists, Bulletin* 66, 1196–1230.

Figure 9.1 Larter, R.D. & Leat, P.T. (2003) *Intra-oceanic subduction systems: tectonic and magmatic processes*. Geological Society, London, Special Publications 219.

Figure 9.2, 9.3 Karig, D.E. (1970) Ridges and basins of the Tonga-Kermadec island arc system. *Journal of geophysical research* 75, (2), 239–254.

Figure 9.4 Taira, A., Ohara, S.R., Wallis, A., Ishiwatari, A. & Iryu, Y. (2016) Geological evolution of Japan: an overview. In Moreno, T., Wallis, S., Kojima, T & Gibbons, W. *The geology of Japan*. Geological Society, London.

Figure 9.5 Karig, D.E. (1971) Structural history of the Mariana Island arc system. *Geological Society of America, Bulletin* 82, 323–344.

Figure 9.6 Elsasser, W.M. (1971) Sea-floor spreading as thermal convection. *Journal of Geophysical Research* 76, (5), 1101–1112.

Figure 9.7, 9.8, 9.9 Oliver, J., Isacks, B., Barazangi, M. and Mitronovas, W. (1973) Dynamics of the downgoing lithosphere. *Tectonophysics* 19, 133–147.

Figure 9.10 Cross, T.A. and Pilger, R.H. (1982) Controls of subduction geometry, location of magmatic arcs and tectonics of arcs and back-arc regions. *Geological Society of America, Bulletin* 93, 545–562.

Figure 10.1, 10.2, 10.3 Wilson, J.T. (1963) A possible origin of the Hawaiian islands. *Canadian Journal of Physics* 41, 863–870.

Figure 10.4 Wilson, J.T. (1973) Mantle plumes and plate motions. *Tectonophysics* 19, 149–164.

Figure 10.5, 10.6 Morgan, W.J. (1972) Deep mantle convection plumes and plate motions. *American Association of Petroleum Geologists, Bulletin* 56 (2), 203–213.

Figure 10.7 Forsyth, D. and Uyeda, S. (1975) On the relative importance of the driving forces of plate motion. *Royal Astronomical Society, Geophysical Journal* 43, 163–200.

Figure 10.8 McKenzie, D.P. (1969) Speculations on the consequences and causes of plate motions. *Royal Astronomical Society, Geophysical Journal* 18, 1–32.

Figure 11.1, 11.2 Mitchum, R.M., Jr., Vail, P.R. and Thompson, S., III (1977) Seismic stratigraphy and global changes of sea level, Part 2: the depositional sequence as a basic unit for stratigraphic analysis. In Payton, C.E. (ed.) *Seismic stratigraphy – applications to hydrocarbon exploration*. American Association of Petroleum Geologists, Memoir 26, Tulsa, Oklahoma, 53–62.

Figure 11.3 Sloss, L.L, (1963) Sequences in the cratonic interior of North America. *Geological Society of America, Bulletin* 74, 93–113.

Figure 11.4, 11.5 Mitchum, R.M., Jr., Vail, P.R. and Thompson, S., III (1977) Seismic stratigraphy and global changes of sea level, Part 2: the depositional sequence as a basic unit for stratigraphic analysis. In Payton, C.E. (ed.) *Seismic stratigraphy – applications to hydrocarbon exploration*. American Association of Petroleum Geologists, Memoir 26, Tulsa, Oklahoma, 53–62.

Figure 11.8 Miall, A.D. (1997) *The geology of stratigraphic sequences*. Springer, New York.

Figure 11.9, 11.10, 11.11 Vail, P.R., Mitchum, R.M., Jr., and Thompson, S., III (1977) Seismic stratigraphy and global changes of sea level, Part 3: relative changes of sea level from coastal onlap. In Payton, C.E. (ed.) *Seismic stratigraphy – applications to hydrocarbon exploration*. American Association of Petroleum Geologists, Memoir 26, Tulsa, Oklahoma, 63–81.

Figure 12.3 Graham, R.H. 1981. Gravity sliding in the Maritime Alps. From McClay, K.R. & Price, N.J. (eds) *Thrust and nappe tectonics*. Geological Society, London, Special Publication 9, 335–352.

Figure 12.5 Trusheim, E. (1960) Mechanism of salt migration in Germany. *American Association of Petroleum Geologists, Bulletin* 44, 1519–1540.

Figure 12.7 Wernicke, B. (1981) Low-angle normal faults in the Basin and Range Province: nappe tectonics in an extending orogen. *Nature* 291, 645–648.

Figure 12.9B, C Reston, T.J. (2007) The formation of non-volcanic rifted margins by the progressive extension of the lithosphere: the example of the West Iberian margin. From: Karner, G.D.,

Manatscheal, G. & Pinheiro, L.M. (eds) *Imaging, mapping and modeling continental lithosphere extension and breakup*. Geological Society, London, Special Publications, 282, 77–110.

Figure 12.11 Searle, M.P., Elliott, J.R., Phillips, R.J. *et al.* (2011) Crustal-lithospheric structure and continental extrusion of Tibet. *Journal of the Geological Society*, London, 168, 633–672.

Figure 12.12 Searle, M.P., Law, R.D. & Jessup, M.J. (2006) Crustal structure, restoration and evolution of the Greater Himalaya in Nepal-South Tibet: implications for channel flow and ductile extrusion of the middle crust. From: Law, R.D., Searle, M.P. & Godin, L. (eds) *Channel flow, ductile extrusion and exhumation in continental collision zones*. Geological Society of London, Special Publications, 268, 355–378.

Figure 12.14 Dewey, J.F. (1988) Extensional collapse of orogens. *Tectonophysics* 7, 1123–1139.

Figure 12.15 Okay, A.I. (2000) Geology of Turkey: a synopsis, *Anschnitt*, 21, 19–42.

Figure 12.16 Sans de Galdeano, C. (2000) Evolution of Iberia during the Cenozoic with special emphasis on the formation of the Betic Cordillera and its relation with the western Mediterranean. *Ciências da Terra (UNL), Lisboa* 14, 9–24.

Preface

Throughout my lengthy career in geology as student, researcher and teacher, I have experienced many changes in the way geologists have interpreted the natural world, many of them exciting and a few that were genuinely revolutionary. Geological research does not flow steadily onwards by means of small incremental advances, but can be better understood as a series of significant discoveries, or changes in interpretation, that have transformed the way we understand the Earth. Each of these new ideas encouraged a burst of activity as researchers attempted to apply it more widely in order to test its universality, and thereby its validity as a scientific theory. Probably the best example of such a transformative idea is Plate Tectonics, which, although questioned at the time it was introduced, is now universally accepted as a general principle, and a large number of the subsequent advances in geological understanding have been based upon it. This book is a selection of some of these new ideas ('breakthroughs') that I have found particularly interesting. The chapters are arranged in roughly chronological order, depending on when I believe the key ideas were first published, and each represents a different idea, discussed in its historical context.

Any attempt to identify the most significant ideas in the history of geological research is bound to be subjective and controversial. My list is somewhat biased towards my own interests in tectonics and structural geology, and is certainly not exhaustive – many other ideas could have been chosen. In each case I have attempted to summarize the new idea, to contrast it with previous views, and to convey my impression of its impact on geological science. I describe each as far as possible in the form in which it was originally published, using key quotations from the relevant texts, followed by a brief summary of how the early ideas were modified and improved over the decades that came after their introduction.

Each chapter is worthy of a volume of its own if it were to be thoroughly explained with all the relevant references, but in order to keep the book to a reasonable size, I have been quite selective in citing or discussing only those references that seemed to me to be the most relevant. However, I have tried to identify the principal sources of each idea, bearing in mind that, very often, the person or persons credited with the idea may have been influenced by others whose role is not obvious from the published literature.

I hope that readers might share my sense of excitement on first discovering these ideas.

Graham Park, February 2019

Acknowledgements

I am indebted to Professor John Winchester for his many helpful comments and suggestions, to Professor Robert Butler for several improvements to Chapter 8, and to my wife, Sylvia, for her unfailing support and encouragement. Any remaining deficiencies in this book are my own responsibility!

1

Uniformitarianism: the first breakthrough

The credit for the first serious attempt to apply scientific principles to the study of the Earth surely belongs to James Hutton (1726–1797). Hutton is often called 'the Father of modern Geology', not because he was the first to discuss the subject – many philosophers and naturalists had speculated about the origin of the Earth before him – but because he was the first to propose a theory based on the evidence that he had gained from studying rock samples and natural exposures, and then to follow it up by seeking and finding new evidence to support it. His *Theory of the Earth*, first published in 1788, contained two revolutionary concepts: firstly, that all the rock sequences presently exposed on the Earth's surface could be explained by natural processes that could be observed and described – *i.e.* that 'the present is the key to the past'*; and secondly, that these processes had operated over long periods of time (he specifically mentions 'millions' of years) and that 'there was no vestige of a beginning, no prospect of an end'. Both these new concepts, though challenged at the time by many in the religious establishment on the grounds that they diminished the divine role in creation, were eventually accepted by the scientific community and form the basis of modern earth science. These two concepts were subsequently consolidated by Charles Lyell, who popularized and expanded Hutton's views, into the term 'uniformitarianism'.

The background to Hutton

Early developments

Speculations about the origin of the Earth and the nature and physical composition of its surface materials date back at least to the ancient Greeks and almost certainly before them. The recognition that fossiliferous sedimentary strata were of marine origin has been attributed to medieval philosophers in both China and Persia. However, serious attempts to devise a theory about rocks and their origins seem not to have commenced until the late eighteenth and early nineteenth centuries – the term 'geology' (from the Greek: 'earth study') was apparently first defined by De Saussure in 1779. Prior to the work of James Hutton, the prevailing theory concerning Earth's origins was known as 'Neptunism'.

*Phrase attributed to Archibald Geikie (see Chapter 3).

Neptunism

The term 'Neptunism' (after the Roman God of the Sea) is credited to Jean-Etienne Guettard, but the most influential exponent of the concept was Abraham Gottlob Werner (1749–1817). Werner believed that a universal ocean, equated with the biblical Great Flood, was responsible for producing almost all the rocks and minerals now exposed in the Earth's crust (active volcanic extrusions were an exception), and that this ocean had then receded to its present location. This idea led him to the concept of a series of universal layers which could be recognized in every continent on the basis of their composition. The principle of 'superposition' – that younger strata lay above older – had been established by Nicolas Steno in the previous century, and thus Werner envisaged his successive layers as representing a time series, from the oldest at the base upwards:

1. a 'primitive' layer encompassing granites, gneisses and other crystalline rocks;
2. a 'transition' series of (supposedly) unfossiliferous greywackes, slates etc;
3. a 'secondary' series of indurated but stratified and obviously fossiliferous sediments such as limestones;
4. poorly consolidated sands, gravels and clays;
5. young volcanic extrusives obviously associated with current volcanicity.

There were two critical flaws in this theory: the first involved the origin of crystalline igneous rocks, particularly those basalts found interlayered with sedimentary strata. Such rocks were believed by Werner to be formed by chemical precipitation from the universal ocean, and layered basalts were not recognized to be the same rock that was currently produced by volcanoes. The second major problem was the whole concept of the universal ocean: this would have to contain an enormous volume of water in order to submerge the whole Earth up to the summits of the highest mountains. In order for the sea level to be reduced to its present position, this water would have to be removed – to where? Moreover, the poorly consolidated gravels and sands, etc. of Werner's youngest sedimentary layer are found at varying heights; this has somehow to be explained by the receding ocean.

Werner published his main work *Von der äusserlichen Kennzeichen der Fossilien* (on the origin of the external characteristics of fossils) in 1774 and an English edition was published in 1805. (Note that the term 'fossil' as then used included minerals and other non-organic material.)

The contribution of James Hutton

James Hutton was born in Edinburgh in 1726 and studied classics at Edinburgh University. He gave up a short career as a lawyer to concentrate on his experiments in chemistry, and at the age of 18 began to study medicine at Edinburgh University before travelling to Paris to continue his medical studies, finally obtaining a doctorate at Leiden University in 1749. After returning to Edinburgh, he abandoned a career as a physician to continue with his studies in chemistry. He was friendly with other prominent members of the Scottish Enlightenment, including the mathematician John Playfair, the philosopher David Hume and the economist Adam Smith, and formed a close partnership with Joseph Black in his chemical investigations.

In the 1750s, he moved to the family farm in Berwickshire and concentrated on agricultural improvements. By the 1760s, he had formulated many of his theories about the Earth, and in 1764 made a geological tour of northern Scotland. For the next 24 years, Hutton developed his theories, communicating with numerous geologists, and received rock and mineral specimens from many parts of the world. He was familiar with the work of many of the prominent geologists of his day, including Horace-Bénédict de Saussure, Jean-André de Luc and George-Louis Leclerc, Compte de Buffon, who are quoted extensively in his book. Hutton's investigations culminated in 1787 and 1788 with his study of unconformities in Arran and the Southern Uplands of Scotland.

Hutton was the joint founder of the Royal Society of Edinburgh in 1783, and first published his theories in their journal in 1788. His famous *Theory of the Earth* was published in book form in 1795.

Hutton's principal contribution to geology, encapsulated in the concept of uniformitarianism, resulted from his investigations into several different processes: 1, erosion, sedimentation and lithification; 2, volcanism and the origin of igneous rocks; 3, the elevation of mountains; and 4, the formation of unconformities – leading him ultimately to the realization of the immensity of geological time.

Erosion and deposition

Hutton's detailed agricultural observations of soil being washed from the land and deposited into ditches and river beds, together with his knowledge of the rocks around Edinburgh, led him to conclude that sedimentary rocks such as mudstone, sandstone and limestone were formed by the observable processes of erosion of the land surface and the deposition of the derived material into a lake or sea. He extended this observation to conclude that all stratified rocks found on land were originally deposited on the sea bed and that this implied a sequence of events that must have occurred over a significant period of time:

> 'We find the marks of marine animals in the most solid parts of the
> earth; consequently, those solid parts have been formed after the
> ocean was inhabited by those animals which are proper to that fluid

medium. If, therefore, we knew the natural history of these solid parts, and could trace the operations of the globe, by which they have been formed, we would have some means of computing the time through which those species of animals have continued to live.'

The problem of lithification

The Neptunian view was that all the materials that made up the sedimentary rocks were laid down or precipitated from water. Hutton was critical of the Neptunists for not providing any evidence to support this. He noted that fragments of older material such as pebbles, grains and fossils in sedimentary rocks were held together by crystalline material, such as calcite or quartz, which he believed could not be dissolved out of water as the Neptunists claimed. It should be remembered that the composition of water was only determined in 1766. Hutton, from his experience of chemistry, maintained that there was no known example of quartz, calcite or other cementing materials being precipitated from water. He knew that crystalline igneous rocks were precipitated from melts, and his initial response to the problem of lithification was to suggest that indurated sedimentary layers were consolidated into hard rock by being partially melted by means of heat from below, and that this subterranean heat was also responsible for the uplift of the sedimentary strata to their present positions on land. Hutton criticizes various Neptunists, including de Saussure, de Luc and Buffon, at great length in *Theory of the Earth* for not providing evidence to support their theory.

It was only later that Hutton realized that the slow growth of stalactites (Fig. 1.1A) showed that calcite could indeed be deposited from a weakly acidic solution. Then, having obtained specimens of silica from the hot spring deposits at Geysir in Iceland (Fig. 1.1B), he wondered whether the nature of the hot water was responsible, and asked a colleague who was due to visit Iceland to collect some of the volcanic water. This was analysed for him by his friend Joseph Black, who proved that the water was alkaline and contained dissolved silicon. Hutton explains this as follows:

> 'although siliceous substance is not soluble, so far as we know, by simple water, it is soluble by means of alkaline substance; consequently, it is possible that it may be dissolved in the earth…
> [and concludes] My conjecture has thus been justified.'

This is an example of an experiment designed by Hutton to test his hypothesis that silicification in sedimentary rocks could be explained simply by deposition from weakly alkaline solutions. He concluded that alkaline or acidic water contained in sedimentary strata, and heated from below, could dissolve silica or calcite, be driven upwards, and could deposit the soluble substances nearer the

Figure 1.1 A Stalactites in limestone cave, Gadime, Kosovo. Shutterstock, by Kagai19927.
B Strokkur geyser, Iceland, surrounded by hot-spring deposits of silica. © Shutterstock, by Santi Rodriguez.

surface. He thus eliminated the main objection to his earlier views on indurated sediments.

Volcanicity and igneous rocks

Hutton recognized the critical distinction between rocks that have been originally derived by erosion then deposited in the sea and consolidated, and rocks such as granite and basalt that have crystallized from a melt. In this, he diverged fundamentally from the views of the Neptunists. Hutton was familiar with the active volcanoes of Italy, and noted the similarity between the lavas produced from those active volcanoes to the basalts found in Scotland. He deduced that the basalt rocks of Salisbury Crags in Edinburgh (Fig. 1.2) were of volcanic origin. He observed how veins of granite and dykes of basalt penetrated into their host rocks, concluding that they must have been molten, and younger than the host material. He proposed that the interior of the Earth was hot, and that this heat was responsible for the creation of new rock. Adherents of these views were subsequently known as 'Plutonists' in contrast with the Neptunists.

The elevation of mountains

Having travelled to the Alps, and being familiar with the observations of several continental naturalists, Hutton knew that layers of limestone containing the fossil shells of marine creatures, identical to those found in present-day seas, occurred high up in the Alps and other great mountain chains. He realized

Figure 1.2 Salisbury Crags, Edinburgh. Hutton realized that this was the outcrop of an intrusive body of igneous rock, derived from a melt. © Shutterstock, by cristapper.

that some mechanism was required to elevate them from their original site of formation at the bottom of the sea to their present position. He was also aware of the widespread occurrence of igneous rocks such as granite in mountainous terrains, and knew that great heat was required to melt them. He therefore suggested that 'subterraneous fire' – a deep-seated heat source – caused the expansion necessary to elevate the mountains, arguing that if volcanoes such as Etna and Vesuvius could produce sizeable mountains purely as a result of internal heat, then surely the same process could also account for the much greater expansion necessary to explain the Alps. He noted that volcanoes occurred in many places around the margins of the Alps but not within the mountain chain itself, supposing that, if the subterranean heat could not escape via molten lava from volcanoes, it may have provided a powerful mechanism for elevating the chain. He also believed that the intense 'contortions' (i.e. folding) of the strata observed in the Alps could be explained as part of the process of uplift of the massif:

> 'The strata of the globe are actually found in every possible position.
> For, from horizontal, they are broken and separated in every
> possible direction, and from a plane, they are bent and doubled.
> It is impossible that they could have originally been formed, by
> the known laws of nature, in their present state and position; and
> the power that has been necessarily required for their change, has
> not been inferior to that which might have been required for their
> elevation from the place in which they had been formed.'

The significance of unconformities

The key role of unconformities in guiding Hutton's thinking about geological time is evident from his description of several examples that he studied in 1787 and 1788. The first was on the island of Arran in 1787, but the following year he found better examples of an unconformity near Jedburgh, and by following the likely route of its outcrop across the Southern Uplands, he finally came upon the most famous example on the coast at Siccar Point, in Berwickshire (Fig. 1.3). From these examples, he observes that the older strata (in this case, of Lower Palaeozoic age) had been uplifted, tilted and eroded before being sunk beneath the sea in order for younger (Old Red Sandstone) strata to be deposited on top. He describes 'puddingstone' (i.e. conglomerate), interposed between the two sets of strata, composed of rounded fragments of the older series. He concludes:

> 'It is plain that the schisti [i.e. cleaved sandstones and shales] had
> been indurated, elevated, broken, and worn by attrition in water,
> before the secondary strata, which form the most fertile parts of

Figure 1.3 Hutton's unconformity at Siccar Point, Berwickshire. Hutton deduced from this outcrop a sequence of events: 1, deposition of older series; 2, tilting into a vertical position; 3, erosion of older series; 4, deposition of younger series. © Shutterstock, by Mark Godden.

our earth, had existed. It is also certain that the tops of our schistus mountains [i.e. the Southern Uplands] had been in the bottom of the sea at a time when our secondary strata had begun to be formed.'

Hutton explained the differences between the mountainous and lowland regions of Scotland in terms of their different age and degree of induration:

'If we shall thus allow the principle of consolidation, consequently also of induration, to have been much exerted upon the strata of the alpine [i.e. mountainous] country, and but moderately or little upon the strata of the low country of Scotland, we shall evidently see one reason, perhaps the only one, for the lesser elevation of the one country above the level of the sea, than the other.'

From these observations on unconformities, Hutton produced a simplified tripartite division of the geology of Scotland into: 1, the highly indurated and often steeply inclined 'Alpine', or elevated country (e.g. the Southern Uplands and Highlands); 2, the gently tilted softer strata of the flat or lowland country (the Midland Valley and Moray Firth basin) and 3, various mountains, dykes and veins of igneous material that is evidently younger than the strata of the lowlands.

Geological time

Since the sedimentary strata on the continents are clearly the products of erosion, Hutton concludes that a previous landmass must have existed in order to provide the materials required:

'we have reason to conclude that, during the time this land was forming, by the collection of its materials at the bottom of the sea, there had been a former land containing materials similar to those which we find at present in examining the earth.'

In considering the time necessary to explain his interpreted sequence of geological events, Hutton compares the present coastlines of the Mediterranean with those described by the ancient Greeks and Romans and found no evidence of significant changes. As these processes were obviously very gradual, he emphasizes that great timespans are necessary to explain the known geological record:

'all the coasts of the present continents are wasted by the sea, and constantly wearing away upon the whole; but this operation is so extremely slow, that we cannot find a measure of the quantities in order to form an estimate. So the present continents of the earth, which we consider as in a state of perfection, would, in the natural operations of the globe, require a time indefinite for their destruction.'

He then concludes:

'For having, in the natural history of this earth, seen a succession of worlds, we may conclude that there is a system in nature ... it is vain to look for any thing higher in the origin of the earth. The result, therefore, of this physical inquiry is, that we find no vestige of a beginning, no prospect of an end.'

In terms of both the formation and age of the Earth, this final oft-quoted phrase – 'no vestige of a beginning, no prospect of an end' – was to become the defining statement of the secular view held by the proponents of uniformitarianism in the scientific community, and was in conflict with the contemporary religious orthodoxy, that the Earth was formed in a single event no more than a few thousand years ago.

A theory of the Earth
Hutton's views were first published in the *Transactions of the Royal Society of Edinburgh* in 1788 but did not appear in book form until his *Theory of the Earth* in 1795. It was Hutton, known as the 'Father of modern Geology', who is generally credited with the establishment of geology as an independent science.

Hutton's legacy
Hutton died shortly after the publication of *Theory of the Earth*, in 1797, as the result of a serious illness. Unfortunately his rather convoluted and over-elaborate writing style meant that his ideas did not obtain as much attention as they

Figure 1.4 Portrait of Charles Lyell. Lyell was a supporter of Hutton and admirer of his work. In publicizing Hutton's theories, incorporated in his *Principles of Geology*, Lyell ensured that uniformitarianism became accepted. © Shutterstock, by Everett Historical.

deserved at the time. However, his ideas were popularized both by his friend John Playfair and in two influential textbooks, *Principles of Geology* and *Elements of Geology* by Sir Charles Lyell, both of which ran to numerous editions spanning the period 1830 to 1865.

Lyell (Fig. 1.4), born of a wealthy Scottish family, and of independent means, was appointed Professor of Geology at University College, London and spent much of his time travelling in North America and Europe observing geological phenomena of all kinds. He was a close friend of Charles Darwin and influenced the latter's views on evolution. Lyell's work covered the whole field of geology (the first edition of his *Principles* textbook extended to three volumes) but a major contribution was to make Hutton's views more accessible to the general scientific community.

The main reason that the views of Hutton and Lyle eventually prevailed over those of Werner and the Neptunists was that they were largely based on observations of processes that could be studied (such as sedimentation and volcanism) so that they were more readily subject to scientific testing in a way that Neptunism, being more of an abstract theory, was not.

Catastrophism *vs* uniformitarianism

Catastrophism is the theory that the history of the Earth has been determined by one or more short-lived violent events such as the 'universal flood', and contrasts with the concept of uniformitarianism, expounded by Hutton and Lyell, which holds that slow incremental changes such as erosion and sedimentation are responsible for the Earth's geological features – i.e. all geological processes throughout the past were similar to those that could be observed at the present day ('the present is the key to the past').

Georges Cuvier (1769–1832), often called the 'father of Palaeontology', was the leading exponent of catastrophism during the early part of the nineteenth century. Cuvier was an outstanding French naturalist and zoologist, who produced a large body of detailed work, based on comparative anatomy, comparing fossils to living organisms, which enabled him to integrate fossils into the Linnaean classification. In contrast to much contemporary opinion, which held that there was no distinction between fossils and living organisms, he proved that certain fossils represented animals that had become extinct, and that extinction was permanent and attributable to a catastrophic event. He believed that new fossil forms appeared suddenly in the fossil record and remained unchanged until extinction, at which point they were suddenly replaced by new ones. This idea was, of course, subsequently challenged, both by William Smith (see chapter 11) and by Darwin's views of evolution.

Cuvier's most influential work was *Théorie de la Terre* first published in 1813 (translated into English by Robert Kerr in 1821) in which he established the basic principles of biostratigraphy. Cuvier was a Protestant secularist, product of the French 'enlightenment', and avoided any reference to divine intervention – his catastrophes were extreme natural events. However, some of his followers, such as William Buckland and Robert Jamieson, explicitly linked catastrophism to the biblical account of Noah's flood and attributed the formation of the Earth and the violent events that shaped it to divine intervention, using biblical material in support.

The uniformitarian and gradualist ideas espoused by Hutton and Lyell challenged Cuvier's interpretation of geological cataclysms, and from around the middle of the nineteenth century the views of the former prevailed. However in more recent decades, there has been a recognition that certain catastrophic events have indeed played a part in Earth history, and present-day views are actually an amalgamation of uniformitarianism and catastrophism – a story of gradual slow changes punctuated by occasional catastrophic events such as strong earthquakes and explosive volcanism. The turning point was the suggestion in 1980 that the Chicxulub crater in Mexico was caused by the impact of a giant asteroid, and that the resulting catastrophic environmental changes caused the end-Cretaceous mass extinction. Igneous petrologists have identified a number of volcanic 'super-events' which have produced a series of 'large igneous provinces' in the course of Earth history, each of which would have been many times more catastrophic in its effects than more recent events such as the well-studied 1980 eruption of Mount St. Helens or even the destruction of Krakatoa in 1883.

2

Evolution and the *Origin of Species*

The idea that all animals and plants are descended from ancestral forms that are different (more 'primitive' or less complex), and indeed that all forms of life are based on a common ancestor that lived far back in geological time, is one of the key precepts of the modern geologist. However, when it was first introduced in the nineteenth century, it was truly revolutionary, caused one of the greatest and most acrimonious debates among the scientific community, and is still a very live issue today, especially in the USA.

Historical background

The concept of one type of organism being derived from another dates back to the early Greek philosophers and was supported by the Roman philosopher Lucretius, but most thinkers in the Classical period believed that the nature of all things was fixed and had been established by divine order – a view that subsequently became incorporated into Christian dogma. Only in the seventeenth century was this religious orthodoxy challenged, when some scientists began to explain natural phenomena by physical laws that did not require divine intervention. The biological sciences did not benefit from these changes in thinking at that time; when Carl Linnaeus introduced his biological classification in 1735, he recognized the essentially hierarchical nature of species relationships but believed that the form and nature of each individual species was determined as part of the divine plan. Other eighteenth-century naturalists, including Pierre-Louis Maupertuis, Georges-Louis Leclerc, Comte de Buffon and Erasmus Darwin (Charles Darwin's grandfather), speculated about evolutionary changes, but there was no suggestion of a possible mechanism until the 'transmutational theory' was introduced by Jean-Baptiste Lamarck in 1809.

Lamarck proposed that simple forms of life achieved greater complexity by actively developing anatomical changes, and that organisms inherited changes introduced by the previous generation as a result of the use or disuse of various anatomical features in an attempt to better adapt to environmental pressures (e.g. limbs could grow longer and eyes could see better, merely by using them). However, Lamarck was heavily criticized both by the religious establishment (e.g. by William Paley in 1802 in his *Natural theology or evidence of the existence and attributes of the Deity*) on the basis that his theory diminished the role of God in the creation of the natural world; and also by other naturalists, such as Baron Cuvier (see previous chapter), for not providing any empirical evidence for his ideas.

Another important influence on naturalists of this period was the work of Thomas Malthus in 1798 on the implications of population growth. Malthus pointed out that the inevitable expansion of populations would lead to a 'struggle for existence' in which individuals would compete to acquire increasingly scarce resources. Malthus' work had a particular influence on Charles Darwin.

This, then, was the background to Charles Darwin's *Origin of Species* in 1859.

Charles Darwin

Charles Darwin was born in Shrewsbury in 1809, the son of a prosperous doctor. In 1825, his father sent him to Edinburgh University in the expectation that he would obtain a medical degree, but the young Darwin neglected his studies and took more interest in the natural sciences, including geology. After three years in Edinburgh, his father, despairing of Charles following his own career, sent him to Cambridge to study for a Bachelor of Arts degree, hoping that an alternative career as a country parson might ensue. Instead, Charles spent much of his time pursuing his interest in botany, becoming friendly with John Henslow, then the Professor of Botany, who was later instrumental in obtaining for Darwin the position of Naturalist on the Beagle expedition. On graduating in 1831, Darwin attended the course given by Adam Sedgwick, the Professor of Geology, and that summer undertook a geological mapping expedition with him in Wales. While at Cambridge, Darwin familiarized himself with the views of William Paley, who argued for divine design in explaining diversity and adaptation in nature. He was also influenced by Alexander von Humboldt's account of his scientific discoveries in South America, and by the end of 1831 had become well versed in current views in the fields of botany, zoology and geology – ideally suited to the position that now awaited him.

The *Beagle* expedition

In December 1831, Charles Darwin joined HMS *Beagle* as expedition naturalist, having been recommended to Captain Fitzroy by Professor Henslow. This was to be the defining juncture of Darwin's career. The objective of the expedition was to survey and chart the coast of South America, and while the survey work was being undertaken, Darwin would be sent ashore to study the local biology and geology, and in the course of the next five years, he was to amass a vast amount of data, including numerous specimens entirely new to science. While on board, Darwin took the opportunity to study Charles Lyell's *Principles of Geology* and became familiar with Hutton's concept of uniformitarianism in geology.

Amid the numerous discoveries made by Darwin during the voyage was the finding of the bones of several huge quadrupeds, such as *Megatherium*, the elephant-sized 'giant ground sloth', all quite different from any species still living, and now clearly extinct. These were found in the same sedimentary layer as modern marine shells, indicating that the extinction was relatively recent, but with no accompanying evidence of a catastrophic event to explain it. A second important discovery was his identification of raised beaches as indicators of sea-level changes. Another concerned the origin of coral reefs, which he attributed to the sinking of oceanic islands. These were to be the subject of a scientific paper published on his return.

The highlight of the journey in scientific terms was his visit to the Galapagos Islands (Fig. 2.1). Here he noted that a number of different animal species showed differences not only from their close relatives in Chile but also between their representatives in each of the separate islands. He focused particularly on a group of birds subsequently known as 'Darwin's finches' (although at the time, Darwin referred them to several different groups). Two of these birds are illustrated in Figure 2.2A and B. The differences are mainly in the shape of the beaks, which are adapted to different methods of feeding in the various islands. He also noted that the giant tortoise (Fig. 2.3), which is endemic to the Galapagos and is present in nearly all of the islands, exhibits minor variations that are unique to each island.

Figure 2.1 Panorama showing the volcanic landscape of the Galapagos Islands, from the island of Bartolome. © Shutterstock, by FOTOGRIN.

Figure 2.2 A, B Two of Darwin's finches, showing differences in the size of beak adapted to different modes of life. © Shutterstock: A, by Kjersti Joergensen; B, by Wilfred Marissen.

Figure 2.3 Galapagos giant tortoise. © Shutterstock, by Ryan M Bolton.

Darwin returned to England in 1836, having established his reputation as a leading figure in the scientific community, and set about finding experts to catalogue his collections. He met Charles Lyell, who was impressed with Darwin's geological discoveries, and in 1837 read his first paper to the Geological Society of London – on the evidence for the rising of the South American landmass. It was in this same year that he moved to London and began the task of editing and publishing the enormous volume of scientific work that he had accumulated during his travels on the *Beagle* expedition. He was also thinking about a possible theory that would explain species changes: a sketch in one of his notebooks shows an 'evolutionary tree' as an alternative to Lamarck's model of 'independent lineages'.

Work begins on *Origin of Species*

The following year, Darwin was persuaded to become the Secretary of the Geological Society, in addition to all the work he was already undertaking on his scientific results, and his health began to suffer from the illness that plagued him for the rest of his life. Despite this handicap, he started working on his theory of transmutation of species, seeking information from a variety of sources – farmers, pigeon fanciers, plant breeding experts – as well as from various scientists. He was influenced particularly by the ideas of Thomas Malthus on the effects of population growth, realizing that the 'struggle for existence' implied by Malthusian population expansion would result in the preservation of favourable variations and the destruction of unfavourable ones, leading inevitably to the formation of new species. At this point he was able to state: 'I had at last got a theory by which to work'.

In 1839, at the age of only 30, Darwin became a Fellow of the Royal Society, having published influential papers on both geology and biology. He married his cousin Emma Wedgwood, and moved to his new home in London.

The illness and subsequent death of his dearly loved daughter Annie in 1851 caused him to worry that there might have been some hereditary factor involved – although most of his eight children who survived infancy were in fact healthy. Like so many scientists both before and since, Darwin struggled to reconcile his scientific views with his religious faith, and his grief at Annie's suffering and death led him to turn away from religion, and he stopped attending church.

Origin of Species is completed

During the 20 years since he had first conceived of his theory on the origin of species, Darwin attempted to perfect it, delaying publication because of concerns that there were still important problems for which he had as yet no solution. What finally stimulated him to action was the receipt of a paper in 1858 by Alfred Russel Wallace containing almost identical ideas to his own. Wallace had actually published a paper on *The law which has regulated the introduction of new species* in 1855, but for some reason, Darwin did not regard it as a threat to the priority of his own ideas. Darwin was ill at the time Wallace's second manuscript was received, and so his friends Joseph Hooker and Charles Lyell arranged for a joint presentation to the Linnaean Society entitled: *On the tendency of species to form varieties; and on the perpetuation of varieties and species by natural selection*. The following year, *Origin of Species* was published in London by John Murray (see reference on p.246 for full title).

The *Origin of Species*

Variation under domestication

Darwin's great work commences by noting that, under the artificial conditions of selective breeding and careful husbandry, large numbers of different varieties are produced from a single wild parent:

> 'When we look to the individuals of the same variety or sub-variety of our older cultivated plants and animals, one of the first points that strikes us, is, that they generally differ much more from each other than do the individuals of any one species or variety in a state of nature.'

Darwin believed that many domesticated species that now exhibit extreme variability, such as pigeons, poultry, ducks and horses, are descended from a single wild stock. This greater variability is produced by means of the careful selection by expert breeders of certain useful characteristics, which may not have been particularly useful in the wild condition. Taking domestic pigeons as an example, which Darwin had particularly studied, certain varieties are selected for their beauty and others for their speed of flight, and so on, producing varieties so different that, in Darwin's words:

> 'if shown to an ornithologist, and he were told that they were wild birds, would certainly, I think, be ranked by him as well-defined species...'

Despite some views to the contrary, Darwin, like most naturalists, was convinced that all domestic pigeons were descended from the wild rock dove *Columba livia*. To test this, he crossed two distinct breeds, a white fantail and a uniformly black variety; these produced mottled brown and black progeny, which he crossed and produced one beautiful 'grandchild' indistinguishable from the wild rock dove.

Some domestic variation is attributable to changes in the environment under artificial conditions. For example, domestic geese have acquired heavier leg bones and lighter wing bones compared with their wild ancestors because there has ceased to be a need for flight from danger and selection has concentrated on more desirable features. Moreover, domestic conditions are so carefully controlled via special food, shelter etc., compared to conditions in the wild, that certain qualities are inevitably favoured (e.g. heavier udders in cows) rather than others that might be more relevant in the wild state. Darwin believed that much of the variation in domestic species was inheritable and that selection for breeding purposes had been carried out by man since the very earliest times.

Darwin concludes:

> 'The key is man's power of accumulative selection: nature gives successive variations, man adds them up in certain directions useful to him. In this sense he may be said to make for himself useful breeds.'

Variation under nature

Since such great variation can be produced by mankind by selecting only the external and visible characteristics, Darwin suggests that nature would be able to act on 'the whole machinery of life' and would have the whole vastness of geological time to work with. He points out that we see nothing of the slow progression of changes: the fossil record only shows that the various forms of life are different from what they were.

Darwin discusses the difficulty in deciding, in the case of many wild species, whether differences are attributable to varieties of the same species, or to different species, and noted that many authorities disagree on individual cases. He states, however, that:

> 'in the case of animals which unite for each birth and are highly locomotive, varieties rarely exist within the same country but are common in separate but nearby countries.'

This was shown especially in the case of the Galapagos Islands, where each separate island hosted different varieties of particular species (or different species of the same genus). From this he concludes that species which could be highly mixed tended to be more uniform, but that, when geographically isolated, could diverge to the extent that they could be considered different varieties or even different species, and that:

> 'No clear distinction can be drawn between species and sub-species or between sub-species and varieties. These differences blend into each other in an insensible series...'

and further, that:

> '...species are only strongly marked varieties.'

The struggle for existence

The high rate at which all organic beings tend to increase (following the Malthusian principle) produces more individuals than can possibly survive, leading to competition both between individuals of the same species, and between individuals of different species, for food or light, or to find a mate, or escape predation, etc. Darwin notes that these conditions do not apply in cases where alien species are introduced to countries where there are no native competitors or predators, such as the introduction of the rabbit to Australia, where it became common in a few decades. In such cases, in a sense the exception proves the rule.

Darwin notes further that the struggle will almost invariably be most severe between individuals of the same species, since they frequent the same areas, require the same food, and are exposed to the same dangers. This implies that:

> 'the structure of every organic being is related, in the most essential yet often hidden manner, to that of all other organic beings with which it comes into competition for food or residence, or from which it has to escape, or on which it preys. This is obvious in the structure of the teeth and talons of the tiger; and that of the legs and claws of the parasite which clings to the tiger's body.'

Natural selection

Darwin next turns to the problem of how varieties become ultimately converted into distinct species, concluding that:

> 'All these results ... follow inevitably from the struggle for life. Owing to this struggle ... any variation, however slight and from whatever cause proceeding, if it be in any degree profitable to an individual of any species ... will tend to the preservation of that individual, and will generally be inherited by its offspring. The offspring also, will thus have a better chance of surviving, for, of the many individuals of that species which are periodically born, but a small number can survive. I have called this principle, by which each slight variation, if useful, is preserved, by the term "Natural Selection" [in contrast to (author)] man's power of selection.'

This is the core of Darwin's thesis. The fact that existing species are so well adapted to their own particular circumstances testifies to the success of the selection process. A good example is the use of camouflage to avoid predation: leaf-eating insect species are green, whereas bark-feeders are mottled grey; alpine ptarmigan and mountain hares are white in winter, and so on.

Environmental changes

Darwin goes on to consider the effects on the selection process of physical changes to the environment, suggesting that climatic changes, for example, could favour some species and perhaps even cause others to become extinct.

Also, certain forms could migrate into an area, disturbing the balance of those species already there. In the case of an oceanic island, immigration is restricted, which would tend to favour the preservation of slight modifications to existing species rather than their displacement by incoming species. This was borne out by his observations in the Galapagos Islands, where small modifications in beak size in the finches, and in the shape of the carapace of the giant turtles, were restricted to individual islands and could be explained by differences in the available food or way of life best suited to each island.

He also points out the effects on the population, in a large continental area, of periods of subsidence followed by uplift – a phenomenon that he had noted in his earlier work on sea level in South America. Subsidence would result in the formation of a series of large islands, separated by sea, between which interchange within the population would be restricted. Modifications of the original inhabitants would tend to be different in each island, as random changes were preferentially selected but were restricted to each island, thus producing new varieties. When renewed elevation of the continental area took place, some of the new varieties would be successful and spread, while others would become rarer and ultimately extinct. Those species that were in closest competition with the improved forms would tend to suffer the most.

Sexual selection

A particular problem for the early naturalists was how to explain the existence of sexual reproduction. For Darwin, taking its existence for granted, sexual selection was a special type of natural selection, where adaptations were dependent on the struggle between males for the possession of females (hence the production of better weapons such as horns, etc.) or for more beautiful plumage or ornamentation to attract females.

The tree of life

The next section of *Origin of Species* gives us Darwin's ideas on organic evolution, although he does not at this point use that word. He likens the progression of life through geological time to a tree, which offered to the palaeontological community a way of explaining the fossil record. Figure 2.4 is a simplified version of part of Darwin's 'tree of life', where he visualizes a series of time intervals, represented by the horizontal red lines on the diagram, each representing 1000 generations. Starting with 10 species (A–L) of a particular genus, at each successive stage several varieties are produced, some of which are unsuccessful (the thin black lines) and die out before the next stage, while one, or in some cases two, are successful and progress to the next stage. The thick blue and orange lines diverge immediately from species A and continue

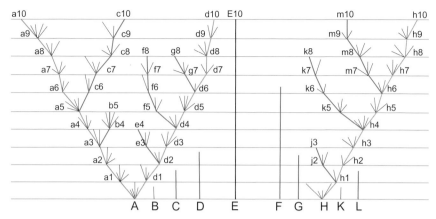

Figure 2.4 'The tree of life'. Darwin's model of evolutionary descent. See text for explanation. Simplified from Darwin, 1859.

as two different series a1–a10 and d1–d10. Further branching occurs at stages a3 and a5 of the blue line, and continue as separate green lines, one ending at b5 and the second continuing to c10. Similar branching takes place from the orange lineage at stages 2, 4 and 6. The purple lineage h1–h10 branches in a similar way from its origin at H. At the eighth generation, there are five successful descendants of A; at stage 10, only three. Altogether, from the original ten species, seven have become extinct at various stages; two, A and H, have produced five successful variants at stage 10 which, depending on the rate of change with time, might be considered to be so different both from their common parents and from each other, as to be considered as separate species. Species E has continued without any change to E10 and represents the case of a 'primitive' form that survives unchanged for long periods of geological time.

At each stage, Darwin suggests that it is either the most extreme of the variants, or the one in the middle of the range, that is successful. At those stages that branch into two successful lines, it is the extreme forms that continue. The species B, C, D, F, G, K and L, which progress to various stages, are envisaged to eventually become extinct due to competition with the more successful variants of the orange and purple lines respectively. By analogy, the horizontal lines could also represent successive geological strata. Darwin notes that the analysis can be extended to explain different families, then orders, depending on the amount of time that elapses between the stages. He states:

'all living things' [are] 'united by complex and circuitous lines of affinities into one grand system.'

Darwin's model is likened to a great tree where some limbs develop into great branches that subdivide into smaller branches and then twigs. Of the many twigs that existed when the tree was a mere bush, only two or three are still present as great branches, whereas the others have died off. His revolutionary idea explains why organisms are related to each other, and contrasts with the concept of the linear descent of organisms, held by some naturalists, where one species is merely replaced by another in a direct line. Moreover, if each species were independently created, as the creationists believed, there is no explanation for the clustering of species and genera into related groups, which Darwin's model so elegantly portrays.

Objections

One of the later chapters deals with Darwin's response to some of his critics. One difficulty that had been raised concerned the absence of transitional forms linking two species, one of which had been supposedly descended from the other. Darwin's theory assumes that both the parent species and all the transitional forms have been eliminated by natural selection, on the assumption that the species that has survived has in some way been able to out-compete all its competitor relatives. Another point that he makes is that the geological record is 'incomparably less perfect than is generally supposed' and that fossils are only preserved under rare conditions:

'The crust of the earth is a vast museum; but the natural collections have been made only at intervals of time immensely remote.'

A related objection was based on the difficulty of some critics to imagine how, say, fish could possibly give rise through natural selection to animals so different as reptiles and birds. Darwin counters this by pointing to flying fish, which have developed the fin to become a primitive wing, and imagining that they could ultimately develop into a winged animal whose oceanic origins might be completely obscured. Speculating on the difficulty of imagining how a land mammal could end up with an entirely marine existence, such as the whale, he points to the fact that some bears swim and catch fish with open jaws, in an analogous manner to certain whale species, and could theoretically develop into separate marine species by selection.

A second problem raised by some critics was that of the origin of certain organs of 'extreme perfection and complication'. The most often quoted of these is the vertebrate eye: the objection being that such an organ was so complex that it had to be 'created' as a whole rather than by stages. Darwin counters this by pointing out that many types of simpler light-sensitive organs exist in nature, and that he had no difficulty in believing that natural selection

'has converted the simple apparatus of an optic nerve merely coated with pigment and invested by transparent membrane into an optical instrument as perfect as is possessed by any member of the great Articulate class.'

This same argument has been much more recently developed in detail by Richard Dawkins in his book *The blind watchmaker*. The important point here is that every small improvement from an initially very simple structure would confer an advantage to the individual and be passed on through natural selection.

One of the most important criticisms, however, came from those who believed that each organic being, both living and in fossil form, had been independently created by God. Darwin counters this by arguing that, if that were so, there is no logical reason why certain species are closely related in form to each other, or to their fossil relatives. Nor does divine creation explain the presence of vestigial organs that appear to have no function, such as the tail in many mammals. Darwin explains this as the product of an ancestral usefulness: mammals are the eventual descendants of fish, in which the tail is obviously of great importance in locomotion, whereas in mammals, natural selection has ensured that it would either be present in a vestigial form (as it is in the *Homo* genus) or be converted for another function, such as a fly-swat in cattle.

Evolution

The final paragraph of *Origin of Species* sees the first mention of the word 'evolution':

> 'There is grandeur in this view of life, with its several powers, having been originally breathed into a few forms or into one; and that, while this planet has gone cycling on according to the fixed laws of gravity, from so simple a beginning endless forms most beautiful and most wonderful have been, and are being, *evolved*.' [My italics]

The response to *Origin of Species*

The reaction to Darwin's masterpiece was mixed. To the geological community, the idea of organic evolution fitted well with the Huttonian views, championed by Charles Lyell, of uniformitarianism and the immensity of time. Many palaeontologists welcomed it as a framework into which the fossil record could be fitted and explained. However, serious objections were raised, on religious grounds, by those who believed that the whole of nature was divinely created and controlled, and that Darwin's views diminished the role of God. Rather than each individual organism being separately created, divine responsibility was to be relegated, at best, to a period in the remote past when life itself commenced – and at worst removed completely.

Figure 2.6 Cartoon depicting Darwin with the body of an ape. By Unknown, Hornet Magazine, via University College, London, digital (Public Domain).

Figure 2.5 Portrait of Charles Darwin in later life. © Shutterstock, by Everett Historical.

Darwin's later career

In his later work, despite continuing ill health, Darwin published *The descent of Man* in 1871. This was one of the most controversial parts of his theory, which had been deliberately omitted from *Origin of Species*. As might have been anticipated, it attracted widespread attention from the public. By then Darwin had become a world-famous and distinguished figure, and had cultivated an imposing beard (Fig. 2.5). However, many commentators were particularly disturbed by the idea that the human race itself was descended from the apes, which provoked a mixture of outrage and amusement (e.g. see the cartoon in Figure 2.6). In the Judaeo-Christian tradition, there was something uniquely special about human beings that made their relationship to the apes particularly objectionable.

Darwin's final publication, in 1881, was on the effects of earthworms on soil. He died in 1882 of heart failure, and was buried at Westminster Abbey, fittingly, alongside Isaac Newton. He is generally acknowledged to be one of the greatest British scientists of all time, and one of a very few who are as widely known by the general public as he is by the scientific community. A marble statue of him sits prominently in the British Museum of Natural History (Fig. 2.7).

Figure 2.7 Statue of Darwin in the Natural History Museum, London. © Shutterstock, by LuisPinaPhotography.

Beyond Darwin

Origin of Species was an immediate success and quickly sold out. Darwin received enthusiastic support from several prominent scientists such as Charles Lyell, Joseph Hooker and Thomas Huxley, but his ideas were opposed by his old friends Henslow and Sedgwick. Huxley famously remarked that 'he would rather be descended from an ape than a man who misuses his gifts'.

However, the main scientific drawback to Darwin's theory of evolution was that, although natural selection was clearly a viable and convincing mechanism for evolution by means of the continuous selection and inheritance of small improvements, the theory provided no answer to the problem of how the variations themselves could be produced. Thus it was possible to argue that all possible variations in structure must already be present in order for the best to be selected. This leads to the absurd conclusion that, for example, the possibility of all existing vertebrate species must have been somehow contained within the eggs of the first vertebrates, and so on, back to the origin of life. Darwin had acknowledged the problem in *Origin of Species*, but was unable to solve it. He suspected that the causes of variation must be held in the male and female 'sexual elements' (i.e. the sperm and the egg) before sexual union takes place. He also thought, contrary to Lamarckian theory, that differences (e.g. of climate, food etc.) were unlikely to produce species variation directly, on the grounds that variations appeared at birth, or at very early stages, when environmental factors could not have been brought to bear. This problem was not to be finally resolved until the mechanism of genetic mutation was discovered in the twentieth century.

Mendelian inheritance

Support for Darwin's thesis was provided in 1866 by Gregor Mendel, who showed through numerous experiments that where two parents exhibited different characteristics, these could be inherited in a systematic and predictable way. A well-known example is his experiments with peas. Using two true-breeding varieties, a short one and a tall one, he found that if he cross-bred them, the first generation were all of the tall variety. However, of the second generation (i.e. the grandchildren) three-fourths were tall and one-quarter were short. He deduced that shortness and tallness were due to inheritable 'factors' (later to be called 'genes') of which the tall factor was the dominant and the small the recessive. The recessive factor was in some way present but hidden in the second-generation plants.

Genetic mutation

In 1901, unaware of Mendel's work at the time, the Dutch botanist Hugo de Vries suggested that the inheritable characteristics were transferred by 'particles', for which he coined the name 'pangenes' – subsequently changed to 'genes' – and that evolution took place by means of sudden inheritable changes in the genes, to which he gave the name 'mutations'. Watson and Crick provided the final piece in the puzzle in 1953 with the discovery of DNA. It has been shown that mutations can be caused by a large number of factors, including spontaneous molecular decay, exposure to injurious chemicals and various types of radiation, including ultraviolet light and, notoriously, radioactivity.

3

Continental drift

Historical background

Hutton and Lyell (see chapter 1) believed that the elevation of mountains was due fundamentally to vertical processes caused by the internal heat of the Earth and that the folding of strata observed in the Alps and other mountain ranges was a secondary process caused by their elevation.

However, by the later part of the nineteenth century, most geologists who had considered the question of the origin of fold-mountain belts believed that horizontal shortening was responsible, and that this was the result of the shrinking of the Earth's outer layer due to internal cooling. This idea was popular before the publication of Alfred Wegener's revolutionary theory of continental drift, and continued to be popular throughout the first half of the twentieth century until it was finally laid to rest by the accumulation of evidence on oceanic structure and palaeomagnetism, which led in turn to plate tectonics.

The contracting Earth theory

The basis of the contracting Earth theory was the belief that the Earth was slowly cooling from its originally hot molten state, and that the cooling of the molten interior gave rise to a shrinking of the solid outer shell or 'crust' – rather like the wrinkled skin of a dried-up apple! The eventual discovery of radioactive decay, and the consequent realization that this additional heat source provided an ongoing supply of heat, meant that the cooling Earth theory had eventually to be abandoned, but this was not generally acknowledged until the mid-twentieth century.

It is not easy to determine the true originator of the contracting Earth theory. It has been attributed to the American scientist James Dwight Dana (1813–1895) but the theory was embraced by many other prominent geologists of the period, including Elie de Beaumont and Eduard Suess. In his 1982 survey of classical theories of orogenesis, Celâl Şengör states that de Beaumont had the prior claim to the theory.

Elie de Beaumont

In 1852, French geologist Elie de Beaumont (1798–1874) in his *Notice sur le système des montagnes* suggested that the worldwide system of fold-mountain ranges was produced by the lateral compression caused by the gradual contraction of a cooling Earth. He also made the significant observation that the age of a mountain system could be determined by the ages of the youngest deformed strata and the oldest undeformed strata. However, a much more comprehensive treatment was subsequently provided by the American geologist James Dwight Dana.

James Dana

James Dwight Dana (1813–1895) was the Professor of Geology at Yale University and was familiar with recent field work in the Appalachian fold-mountain system. In his *Manual of Geology*, published in 1863, the following passage in the section entitled 'Changes of temperature producing expansion and contraction' gives a clear exposition of the theory. He states that a change of temperature may have acted:

> 'by contraction going on within the Earth's interior beneath its solidified crust. The fact that this cause has acted in the Earth's past is beyond question if the globe was once in a fused state, as is generally believed by geologists. Since the crust when formed would have the size which the globe at that time had, all subsequent cooling, as it would tend to diminish the interior, would bring a slowly increasing strain upon it, and, unable to accommodate itself to the changing size by any process of shrinkage, it must do so either by fractures or plications or both.'

He goes on to suggest that the inequalities of thickness and texture across large areas of the Earth's crust would control the location and distribution of mountain belts.

In North America generally, because of the obvious regularity of the fold pattern in the Appalachian Mountains, Dana's ideas proved popular. Another exponent of the contractionist theory was the highly respected Eduard Suess, who at that time had an unrivalled knowledge of the geological structure of the Western Alps.

Eduard Suess

Suess (1831–1914) was the Professor of palaeontology and, subsequently, geology in the University of Vienna. In his 1875 work, *Die Enstehung der Alpen*, Suess rejects the idea, held by Hutton and others, that mountain chains such as the Alps were formed by vertical uplift caused by a rising crystalline core, and states that his own observations indicated that horizontal motion was much more important, ascribing the source of this horizontal motion to global contraction.

The importance of horizontal movements in the generation of fold and fault structures had become more apparent to several of those geologists who had direct experience of mountain belts such as the Alps, and in 1884, Marcel Bertrand re-interpreted the famous Glarus double fold as a single northward-directed 'nappe' (Fig. 3.1), which implied 40 km of horizontal movement. This provided significant support for the contractionist theory.

In Britain, work by the Geological Survey in Northwest Scotland in the late nineteenth century was providing further evidence of the importance of horizontal movements. Archibald Geikie's widely used textbook affords a good example of how the theory was viewed at that time.

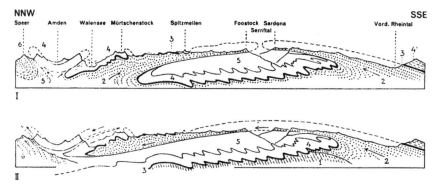

Figure 3.1 The Glarus double-fold: 1, according to Escher von der Linth and Heim; and II, according to Marcel Bertrand. From Şengör, 1979.

Archibald Geikie

Sir Archibald Geikie (1835–1924) was appointed as the first Professor of Geology at the University of Edinburgh in 1871 and subsequently as Director-General of the Geological Survey of Great Britain from 1881 to 1901; his *Textbook of Geology* was the standard guide for British geology students for many decades. His views on mountain formation can be summed up in the following quotations from the third edition of his book, published in 1893:

> '...the true mountain ranges of the globe...may be looked on as the crests of the great waves into which the crust of the Earth has been thrown.'

And:

> 'These examples [i.e. the Alps, Rockies etc. (author)] show that the elevation of mountains, like that of continents, has been occasional

and sometimes paroxysmal. Long intervals elapsed, when a slow subsidence took place, but at last a period was reached when the descending crust, unable to withstand the accumulated lateral pressure, was forced to find relief by rising into mountain ridges.'

And again:

'Geologists are now generally agreed that it is mainly to the effects of the secular contraction of our planet that the deformation and dislocation of the terrestrial crust are to be traced. The cool outer shell has sunk down upon the more rapidly contracting hot nucleus and the enormous lateral compression thereby produced has thrown the crust into undulations, and even into the most complicated corrugations.'

Continental drift

Continental 'drift' was the name given to the concept of the relative movement of the continents around the Earth's surface, and was the first theory that offered the realistic possibility of explaining the localized contractions responsible for the creation of mountain belts such as the Alps. The contracting Earth theory advanced by geologists such as Dana and Suess had eventually to be abandoned once it was generally accepted that the Earth was continually generating heat from radioactive sources, was not cooling down, and therefore not shrinking. However, the contracting Earth theory continued to be upheld by many prominent geologists several decades after continental drift was available as an alternative. The reasons why continental drift took such a long time to be accepted as a valid hypothesis by the scientific community as a whole offer an interesting insight into the background and personalities of some of the key individuals in a debate that lasted for over forty years.

The idea of the relative movement of continents was proposed by the American geologist F.B. Taylor in a paper read to the Geological Society of America in December 1908 and subsequently published in 1910. Taylor proposed that the uplift of mountain belts required enormous forces that could only be explained by continental collision, and that this could account for the shapes of the Alpine–Himalayan mountain chain – notably the convergence of India and Africa with Eurasia. He suggested that the continents could move across the ocean floor by a process of 'crustal creep', and in a later paper proposed that the tidal pull of the Moon could provide the necessary force, thereby undermining the credibility of his theory. The lack of a believable mechanism for the movements, and of any convincing geological evidence in support of his ideas, meant that they received little recognition in comparison with Wegener's.

Alfred Wegener

The theory of continental drift is usually attributed to Alfred Wegener, although some sources refer to the theory as the 'Taylor-Wegener theory'. Wegener himself credits Taylor with priority although he was unaware of Taylor's work at the time of the former's first publication in 1912. In later editions of his book, Wegener cites a number of previous writers who had suggested the possibility of the relative movement of continental masses, generally in the vaguest terms, and discusses Taylor's role, somewhat disparagingly, in the following passage:

> 'I received the impression when reading Taylor that his main objective was to find a formative principle for the arrangement of the large mountain chains and believed this to be found in the drift of lands from the polar regions...'

However, there is no doubt that Wegener produced a far more comprehensive theory with numerous and varied strands of evidence, and it was his work that stimulated the worldwide search for supporting evidence.

Wegener (1880–1930) was a distinguished German meteorologist who had undertaken four expeditions to Greenland to undertake meteorological investigations, during the last of which he tragically lost his life while undertaking a traverse of the ice cap. He published his views on continental drift (in German) in 1912 in the scientific journal *Geologische Rundschau* and in book form in 1915,* to explain the numerous geometric and geological similarities between continents that are now separated by oceans.

The continents of South America, Africa, India, Australia and Antarctica were shown to fit together in a supercontinent called Gondwanaland (originally named by Eduard Suess and now usually referred to as 'Gondwana'). North America and Eurasia were joined in a second supercontinent called Laurasia (Fig. 3.2). These two supercontinents were linked in Central America to form a continuous worldwide landmass termed Pangaea (pronounced 'Pan-jee-a').

Wegener divides the evidence supporting his ideas into four main strands: geophysical, geological, palaeontological and palaeoclimatic.

Geophysical evidence

Wegener's first line of argument uses the repeated longitude measurements then available, which apparently showed changes in the distance between northeast Greenland and Europe of 420 m between 1823 and 1870, and 1190 metres between 1870 and 1907, implying that the current rates of relative movement between North America and Europe (i.e. increases in separation) were between 9 and 32 metres per year. However, the methods then used were subject to large

*The English translation was not published until 1924.

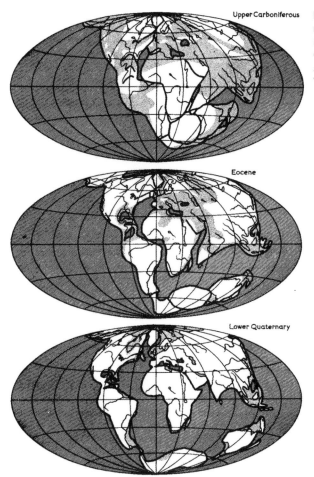

Figure 3.2 Wegener's reconstruction of Pangaea in the Upper Carboniferous, Eocene and Quaternary. From Wegener, 1966.

Upper Carboniferous

Eocene

Lower Quaternary

errors and have been superseded by much more accurate figures of the order of several centimetres per annum from sea-floor magnetic data.

He proposes much more convincing arguments based on isostasy theory, which states that large mountain masses must be underlain by less dense material at depth to provide general gravitational equilibrium. In order for this condition to hold, there must be the possibility of viscous flow of dense material at depth (Fig. 3.3). Wegener believed this denser material (which he terms 'sima') also formed the oceanic crust, and cites seismic surface-wave and magnetic data as evidence that this material had different properties to those of the continental crust. He uses the well-documented example of the post-glacial rise of Scandinavia (Fig. 3.4), which caused an uplift of up to 250 metres over a period of *c*.10,000 years, as proof that if viscous flow beneath the continents could produce movements of this magnitude, then horizontal movements must be equally possible.

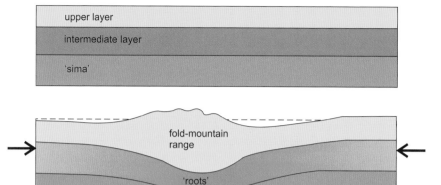

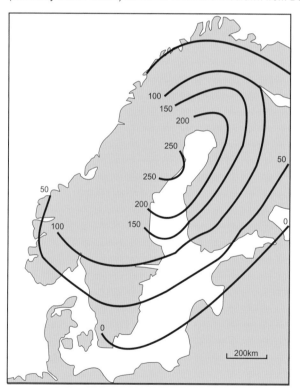

Figure 3.3 Du Toit's explanation of isostasy. A mountain belt formed as a result of the compression and thickening of the crust develops a 'root' by depression of the base of the crust in order to maintain gravitational equilibrium. This can only happen because of viscous flow in a lower layer – the *sima*. The sima corresponds to what we would now call the mantle. Du Toit envisages the intermediate (i.e. lower crustal) layer becoming denser (shown by darker brown) within the root zone. Redrawn from Du Toit, 1937.

Figure 3.4 Post-glacial uplift of Scandinavia. Contours in metres show the height of the post-glacial raised beach above present sea level. The ground surface was depressed by the weight of the ice sheet, and when the ice was removed, the land surface was gradually restored to its previous level. The maximum uplift corresponded to where the ice was thickest – around the head of the Gulf of Bothnia. Wegener used this example to demonstrate the possibility of viscous flow in the substratum beneath the crust. Redrawn from the original diagram by Zeuner, 1958.

Geological evidence

Much of Wegener's argument in favour of continental drift is based on the comparison of the geological structures on each side of the oceans that separate America from Africa and Europe, Africa from India, and India from Australia. Much of the detailed geological data in the later editions of his book came as the result of the work of Du Toit (see below), who was an enthusiastic supporter, and set out to provide as much evidence as he could to back up Wegener's theory.

In the case of the Atlantic, Wegener states:

'Comparing the geological structure of both sides of the Atlantic gives a clear-cut test of the theory that the ocean is an enormously widened rift whose edges were once directly connected. One would expect that folds and other formations that arose before the split occurred would conform on both sides, and in fact their terminal sections on either side…appear as direct continuations of each other.'

Four main lines of evidence are discussed (Fig. 3.5):

1 The Devonian–Carboniferous fold belt (the Cape fold belt) that traverses the southern tip of Africa continues across South America south of Buenos Aires.

2 Gneisses and granites of the Serra do Mar in Brazil match those of Southwest Africa (now Namibia).

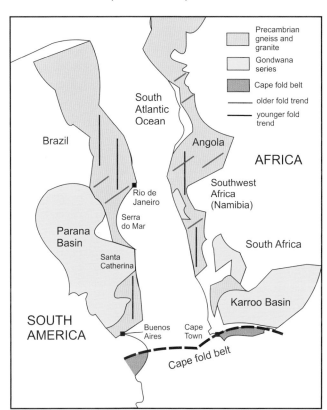

Figure 3.5
Simplified part of Du Toit's map used by Wegener to show the similarities in geology between South America and Southern Africa. The Devonian–Carboniferous 'Gondwana' succession is undeformed in the Parana and Karroo basins but folded in the Cape fold belt to the south. The coastlines of South America and Africa are restored to the pre-drift positions according to Du Toit.

3 Jurassic volcanics of the Santa Catherina system in Brazil are equivalent to those of the South African Karoo Basin.

4 The 'ancient' folds of the (Precambrian) gneiss complexes in both Africa and Brazil possess the same trend pattern: an older NE–SW and a younger N–S, parallel to the coast.

In contrast to these correspondences, Wegener notes that the Atlas ranges of North Africa, which were folded during the Oligocene Epoch (i.e. after the separation of Africa and South America [refer to the Appendix for the geological time periods]) have no counterpart in South America, suggesting that the separation must have occurred prior to the Oligocene.

He notes further that, in the North Atlantic, the Hercynian (Carboniferous) folds of the Appalachian Mountain belt 'extend' into SW Ireland and Brittany (NW France); the Caledonian mountain belt of Norway and the northern British Isles corresponds to the Appalachian belt of Canada (Fig. 3.6); and the older Precambrian gneiss complex of the Hebrides (i.e. the Lewisian) is equivalent to the 'Algonkian' rocks of Labrador.

The mid-Atlantic ridge is explained as a piece of (possibly continental) material left behind when rifts opened up on either side. Iceland and the Azores are considered to be continental blocks formed in the same way.

Wegener goes on to discuss the relationship between Madagascar and East Africa, noting the similarity of the NE-trending gneisses in both places. The remarkably straight coastlines of eastern Madagascar and western India suggested to him that the movement of Madagascar away from Africa may have been parallel to these coasts (i.e. partly strike-slip). The impossibility of the hippopotamus, found in both East Africa and Madagascar, now separated by c.400 km, swimming across to Madagascar is further support for their former contiguity.

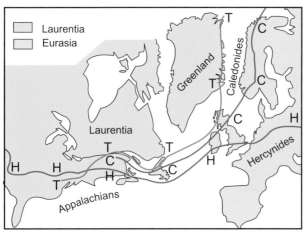

Figure 3.6 Du Toit's map showing the match of the Palaeozoic fold systems across the North Atlantic when restored to their pre-drift positions. T–T, Taconic (late Ordovician) front; C–C, Caledonian front; H–H, Hercynian front. Note that the Caledonian 'front' differs from the one that would be recognized today. Redrawn from Du Toit, 1937.

Turning to India, Wegener notes that the flat gneiss plateau of Western India exhibits the same pattern of older NE and younger N–S fold trends as Madagascar and suggests that the two were once conjoined.

Western Australia is compared with Eastern India in the same way. Folds in the gneiss complex of Eastern India with NE and N–S trends are shared (if suitably rotated) with those of Western Australia. In Eastern Australia, the N–S Carboniferous fold belt corresponds with a similar belt in New Zealand.

In Central Asia, citing Argand's great work *La Tectonique de l'Asie*, Wegener points out that the enormous fold belts of the Himalayas and the Tien Shan require shortening of the Asian continent by around 3000 km, implying that India must originally have been situated by at least that amount to the south of its present position. This was an example of a general pattern in which the front of a moving continent exhibits folding whereas the rear is characterized by tensile fractures and rift structures.

A similar example can be seen in Indonesia, where the twisted shapes of the island arcs surrounding the Banda Sea basin and northeast of New Guinea (e.g. the Banda and Bismarck island arcs) are attributed to distortion caused by the northward movement of the Australia-New Guinea block (Fig. 3.7).

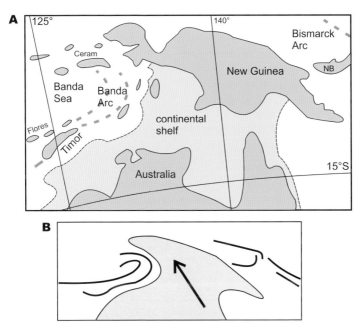

Figure 3.7 Indentation of Indonesia by Australia–New Guinea. **A** Map showing the shape of the continental block of Australia–New Guinea and the contiguous area of eastern Indonesia, including the Banda arc to the north and the Bismarck arc to the northeast. **B** Diagram to illustrate Wegener's interpretation of how the northwesterly movement of Australia has distorted the shape of the adjoining Indonesian island arcs. B, after Wegener, 1924.

Palaeontological–biological evidence

Wegener notes that a significant proportion of the fossil land reptile fauna of Upper Palaeozoic and Mesozoic age is common to both Europe and North America: 64% in the Carboniferous, 32% in the Triassic and 48% in the Jurassic, pointing to a land connection during these periods. Another correspondence is evident in the case of the marsupial fauna of Australia, which is similar to that of South America despite the enormous distance now separating the two continents. In contrast, there are no marsupials in Eurasia, and there has been no exchange of faunas between Australia and the Sunda Archipelago (the nearest part of the Eurasian continent) despite the fact that they are much closer together. There is a similar correspondence in the pre-Cretaceous flora, both between North America and Europe, and between Australia and South America. The similarities in fossil land animals and plants that existed in the different continents in pre-Jurassic time contrast with the obvious differences between the fauna and flora of the separate continents now.

Wegener contends that these coincidences could only be explained by continental drift. The alternative, favoured by many palaeontologists previously, is the former existence of land bridges linking the continents. He rejects this on the grounds that the large volume of continental crust required could not simply disappear. For it to sink to the level of the ocean floor would violate the principle of isostasy, as explained above. He also proposes that the Pacific Ocean must be geologically ancient because it contains many ancient forms of life – for example the cephalopod *Nautilus* (Fig. 3.8) – which are not found in the Atlantic; this fitted his theory of the Atlantic Ocean being much younger (i.e. post-Jurassic).

Figure 3.8 *Nautilus.* This member of the Cephalopod Class of molluscs swims in the Pacific Ocean today, but its ancestors date back to the Cambrian Period, over 500 million years ago. Wegener used its presence as evidence that the Pacific Ocean was more ancient than the other oceans. © Shutterstock, by Mvijit.

Palaeoclimatic evidence

Wegener's arguments based on palaeoclimatic evidence were possibly the most convincing. These include the presence of glacier-derived boulder clays and glacial striations of Permo-Carboniferous age now present in southern Africa, eastern South America, India, and in western, central and eastern Australia; these in their present positions, cover about half the globe (Fig. 3.9) but when restored to their presumed Gondwana fit, correspond to a polar ice cap with a more believable radius of about 30° (compare Figs 3.9 and 3.10). The distribution of other climatic indicators in rocks of the same age also makes sense

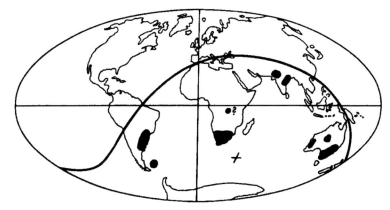

Figure 3.9 Areas of Permo-Carboniferous glacial deposits, together with the position of the equator and south pole (X) that best fit the data. From Wegener, 1924.

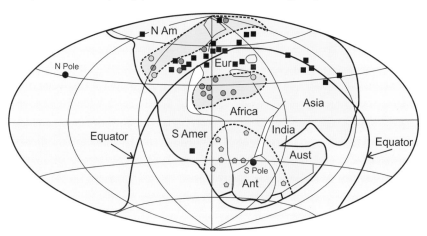

⌂ ice ● evaporite (salt, gypsum) ○ desert sandstone ■ coal

Figure 3.10 Gondwanaland in the Carboniferous – showing climatic belts according to Wegener. Note: 1, the band of coal deposits parallel to the Carboniferous equator (heavy black line); 2, desert areas (in yellow) marked by desert sandstones and evaporite deposits on either side of the equatorial belt; 3, large polar area showing evidence of glaciation. After Wegener, 1924.

when restored to a Gondwana fit; these include dune-bedded sandstones and evaporite deposits (i.e. salt and gypsum), which mark out two desert belts on either side of a central equatorial to sub-tropical rain belt indicated by the presence of coal deposits and coral reefs (Fig. 3.10). The faunal data support this reconstruction: for instance, reptiles only occur in the warmer regions, herbivores in the tropical to sub-tropical belt, and so on. The islands of Svalbard, now within the Arctic Circle, hosted rain-belt flora, including palms and tree ferns during the Jurassic Period, and evaporites in the Carboniferous, while the reverse is true for southern Africa: sub-tropical conditions now but glacial in the Carboniferous.

Many of the detailed geological, palaeontological and palaeoclimatic observations to support Wegener's reconstructions were undertaken after the initial publication of his ideas by Alexander Du Toit (see below) and incorporated in later editions of Wegener's book.

Mountain belts

Wegener believed that mountain belts such as the Alps and Himalayas could be explained as the effect of two continents converging during the amalgamation of Pangaea: Africa with Europe, in the case of the Alps, and India with Asia, in respect of the Himalayas. The oceanic area partly enclosed by the super-continent had been named the 'Tethys Ocean' by Eduard Suess (see above), who believed that the present-day Mediterranean Sea was a remnant of this former ocean, and that much of the former ocean-floor had been incorporated into the Alpine Mountain belt during its formation.

The circum-Pacific belt, which includes the American Cordilleran chain and the Andes, was more difficult to explain by collision, since it required the Pacific Ocean crust to be strong enough to buckle the leading edge of the American continent, which would appear to contradict Wegener's model of a weak ocean crust. Wegener had suggested that the continents were able to move from their original positions in Gondwana and Laurasia to their present ones because the oceanic crust was much weaker than the continents, which could somehow plough their way across it.

A mechanism for continental drift

The source of the forces required to move the continents around the globe proved to be the weak point of Wegener's theory. He considers several possibilities, including 'pohlflucht' (flight from the poles), tidal forces and finally, convection currents.

Pohlflucht had already been proposed as a motive force by several geophysicists as well as by F.B. Taylor, but the consensus appeared to be that it was much too weak, and could be ruled out as a serious contender to explain continental drift. The force is centrifugal, and results from the Earth's rotation, but to calculate its value with any accuracy required knowledge of the viscosity of the substratum over which the continents would have to move – information not available then. In any case, Wegener concludes, following P.S. Epstein, who had attempted the calculation in 1921, that the force would be too small to create the equatorial mountain belts that were an integral part of the drift hypothesis.

Wegener also discusses the possible influence of the force of tidal friction caused by the gravitational attraction of the Sun and Moon. Although this force must exist, and could in theory create lateral movement over geological time, Wegener believed that its current effect would be below the limits of measurement, and that it could not be considered a serious candidate for the motive force in his drift hypothesis.

Schweydar in 1921 had discussed the force produced by the movement of the Earth's rotational axis along a cone-shaped track around a fixed axis (termed 'precession' or 'polar wandering'): this force would be strongest at the Equator, and could theoretically produce a westerly drift. However, the force would affect both the continents and the 'more fluid' material beneath them. Wegener considered that this force could explain, for instance, the opening of the Atlantic but gave it only lukewarm support:

> 'I would like to believe that we have in this deformation of the Earth's
> shape by polar wandering a completely adequate source of the power
> to supply the energy required for folding.'

It should be noted that the term 'polar wandering' subsequently gained a quite different meaning, i.e. the much larger movement of the poles of individual continents through geological time!

Wegener finally comes round to consider what would eventually give the correct solution: convection currents in the 'sima' or basaltic substratum, and quotes Kirsch (1928) as follows:

> 'In conjunction with Joly's idea that the sima under the continental
> blocks is heated by the large radium (i.e. radiogenic) content, and
> that in oceanic regions it cools, Kirsch assumes a circulation – rising
> beneath continents and flowing along under them to the ocean
> regions where it flows downwards, returning to the continents after
> reaching greater depths. Because of the resulting friction, the sima
> tends to disrupt the continental cover and to force the fragments
> apart...'

and concludes:

> 'The forces which displace continents are the same as those which
> produce great fold-mountain ranges. Continental drift, faults and

compressions, earthquakes, volcanicity, transgression cycles and polar wandering are undoubtedly connected causally on a grand scale. Their common intensification in certain periods of the Earth's history shows this to be true. However what is cause and what effect, only the future will unveil …'

The response to Wegener

Wegener's lack of geological background and inability to provide convincing 'geological' evidence to support his theory meant that his ideas received much less support than they deserved, especially in the northern hemisphere. An additional factor affecting his recognition was that his book was initially published (in German) just before the First World War, and was not published in English until 1924. Wegener's views were communicated to a wider audience via a lecture he gave in New York in 1926, but were heavily criticized there. Many geologists, and particularly geophysicists, were influenced by Sir Harold Jeffreys' calculations of the strength of the Earth's crust (see below), and opposed continental drift because they believed that the Earth's crust was too strong to allow the kind of behaviour that Wegener's theory required. However, his ideas were widely accepted among geologists in the southern continents, mainly because of the voluminous supporting work of Alexander du Toit (see below).

Harold Jeffreys

Sir Harold Jeffreys (1891–1989) was a Fellow of St John's College Cambridge, who published highly influential work on mathematics, geophysics and astronomy. His 1924 work, *The Earth, its origin, history and physical constitution* became a standard text on geophysics. Jeffreys was a prominent and effective opponent of the theory of continental drift because he believed that there was no known force strong enough to move the continents across the Earth's surface. His calculations of the strength of the Earth were based on the view that the Earth had cooled from an originally molten state, and that its strength was equivalent to that of the surface rocks. However, work by Sir Arthur Holmes and others in the 1920s (see chapter 4) on the heat generated by radioactive decay had shown that the interior of the Earth was much warmer than Jeffreys had thought – in fact, as we now know, in places close to their melting point at relatively shallow depths of about 50 km. As rock becomes warmer, it also becomes significantly weaker, thus invalidating Jeffreys' calculations.

Jeffreys' views on the origin of mountain belts are well expressed in the following quotation from his later book *Earthquakes and Mountains*, published in 1935:

'The elevation of a mountain system represents work done against gravity... At present the strength of the crust is preventing gravity from making the surface level, but when the mountains were formed it was aiding gravity in resisting the stresses that made them. To explain the origin of mountains we must provide sufficient stresses as will overcome both the strength of the Earth and gravity, for in a symmetrical body, both would act together in opposing any change of shape. The only agency that seems capable of supplying such stresses is contraction of the interior... a sinking crust has to acquire a shorter circumference to fit the new size of the interior.'

The shortening needed was calculated by Jeffreys using the known crushing strength of granite, and indicated that a section of the crust could be compressed by 1/800th of its length before being crushed, and therefore it followed that the Earth's circumference of $c.40,000$ km could be shortened by $c.50$ km before failure (e.g. folding or faulting) occurred. From this he estimated that 70 km shortening would be required to produce the Alps and 190 km for the Himalayas. These figures contrast strikingly with those calculated in modern studies of the actual structures within these mountain belts, which indicate much greater degrees of shortening – around 240 km in the case of the Alps and several thousand kilometres in the Himalayas.

Alexander du Toit

Du Toit (1878–1948) was a South African geologist who had studied mining in London and taught geology at Glasgow University in the early years of the twentieth century, but returned to South Africa and spent many years travelling extensively in southern Africa, South America and Australia gathering the detailed field evidence required to support the geological comparisons necessary to bolster the continental drift theory. Du Toit was an enthusiastic supporter of Wegener, and in 1927 published his studies comparing the geology of southern Africa with South America, and ten years later his comprehensive work *Our Wandering Continents, An Hypothesis of Continental Drifting*, which gave much of the detailed evidence necessary to establish continental drift as a valid theory, to be taken seriously by a large proportion of the geological community.

In addition to providing many further examples of geological matches across the now severed edges of the southern continents, he was able to demonstrate much more convincing evidence for various climatic indicators in rocks of Carboniferous age in the southern Gondwana continents.

It should be noted that the term 'continent' used in a geological sense includes, in addition to the landmass, areas of the adjacent sea bed – the

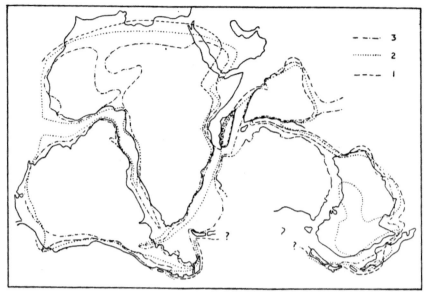

Figure 3.11 Gondwanaland according to Du Toit. Note that this reconstruction includes the continental shelves and is a much more accurate representation of the coastlines than Wegener's. The dotted and dashed lines represent the positions of the shorelines during (1) early Jurassic, (2) early Cretaceous and (3) late Cretaceous time. From Du Toit, 1937, with permission.

continental shelf and continental slope – that are underlain by crust of continental, rather than oceanic, type. Du Toit showed that when the shape of the continents is adjusted to include these, a much better fit of the Gondwana continents is achieved (Fig. 3.11). His reconstruction of the pre-drift continents also differed from Wegener's in that the northern grouping, Laurasia, was centred around the then North Pole and was separated from Gondwana by a wider Tethys Ocean.

Du Toit proved that much of the separation of the Gondwana continents must have taken place during the Mesozoic Era, whereas Wegener had thought that most of the movements had occurred in the Tertiary (Cenozoic) and continued into the Quaternary Period (i.e. the last 2.5 Ma) as well. Du Toit also extended the drift mechanism to explain the older Palaeozoic mountain belts, such as the Caledonian belt, now severed by the Atlantic Ocean.

Support from palaeomagnetism

The Earth's magnetic field can be represented at any point on the surface by a linear vector, which points towards the north magnetic pole, and is inclined at an angle from the horizontal, downwards in the northern hemisphere and upwards in the southern, varying from 0° at the Equator to 90° at the magnetic poles. This magnetic field can be acquired by certain magnetically susceptible rocks such as ironstones and basic igneous rocks when they are first formed and can be recovered once the effects of the present field have been removed.

Work on palaeomagnetism during the 1950s by research groups at Cambridge and Imperial College, London on these 'fossil' fields, when studied for various geological periods in the different continents yielded remarkable results: the apparent magnetic pole positions varied through time, following a path across the Earth's surface known as a 'polar wander curve'. There followed a debate between the exponents of continental drift, who claimed that this proved that the continents had moved through time, and the 'fixists' who explained the data as the result of the movement of the magnetic poles.

Figure 3.12 shows the apparent polar-wander curves (APWs) for each of the five continents, Europe, N. America, S. America, Australia and Africa, as shown by S.K. Runcorn in 1962. It is obvious that these are all quite different, proving that the data could not be explained by the movement path of a single

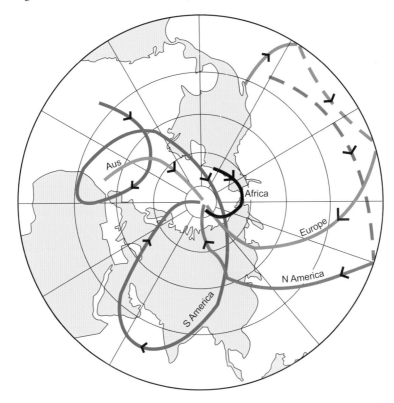

Figure 3.12 Apparent polar wander curves for the continents of Africa (black), Australia (orange), Europe (blue), North America (red) and South America (green) from the beginning of the Mesozoic to the present – ending at the present north magnetic pole. Note the similarity of the European and North American curves over the first part of the period (i.e. during the Mesozoic). The dashed line tracks are in the southern hemisphere. After Runcorn, 1962.

magnetic pole, and that they could only be the result of different drift paths of each continent. Moreover there were some interesting similarities: the paths of the N. American and European APWs, which followed similar tracks through the Mesozoic, were now about 30° from each other, suggesting that the two continents may have been attached during that period. It was realized by Runcorn that, although the positions of the magnetic north pole for Triassic rocks in the two continents now plotted in different places, when these continents were fitted together in the original positions suggested by Wegener, the locations of the European and American poles coincided (Fig. 3.13). This was convincing proof that the continents had drifted to their present positions from their previous locations in the Pangaea supercontinent 200 million years earlier, and finally settled the great continental drift debate.

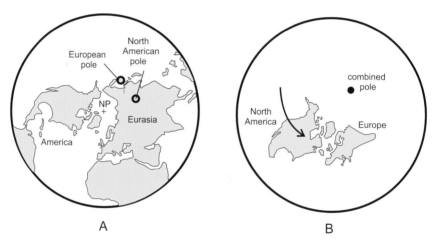

Figure 3.13 Triassic apparent north magnetic pole positions for North America and Europe: **A**, at present; and **B**, with the continents restored to the Pangaea fit of Wegener. NP, present North Pole. After McElhinny, 1973.

Postscript

Wegener's ideas caused considerable, at times acrimonious, debate among the geological community. For a period of over forty years, geologists were divided into 'drifters' and 'fixers', the former being supporters of continental drift, the latter believing that the relative positions of the continents and oceans had been fixed throughout geological time. It was not until the 1950s that advances in palaeomagnetism produced enough evidence to convince geologists that the theory was correct. Wegener was not the first to note the similarity of the fit of the continental margins, or to speculate on the possibility of the continents having changed their relative positions through geological time, but posterity has rightly attributed to him the priority in assembling as complete a theory as the geological facts then available allowed. Moreover, his work was responsible for an enormous volume of useful data as the protagonists of the opposing views strove to prove or disprove his theory.

4

Mantle convection: a mechanism for continental drift?

The weak point in Wegener's theory, and the reason for much of the early criticism of it, was the lack of a credible mechanism for continental drift. The solution to this problem lay in the process of mantle convection – a mechanism first discussed in detail in a brilliant piece of research published in 1929 by Arthur Holmes, one of the most outstanding scientists of his generation. The significance of Holmes' work was realized by Du Toit, and incorporated in the later editions of Wegener's book, as we saw in the previous chapter, but it was not generally accepted by geologists until much later. The key to the understanding of the process of mantle convection was the discovery of radioactivity and its role in temperature distribution, and also, to an extent, greater knowledge of the physical and chemical constitution of the Earth.

Du Toit considered several possible mechanisms for drift, rejecting the contracting Earth model favoured by Suess and Jeffreys, among others, as well as the pohlflucht mechanism. He pointed out that pohlflucht was contradicted by the movements of Gondwana towards the South Pole in the period immediately prior to the Carboniferous, and likewise of Laurentia towards the North Pole immediately afterwards – pohlflucht would obviously imply movements in the opposite direction. He also drew attention to Jeffreys' calculations of the likely magnitude of the pohlflucht force, which ruled it out as too small to be an explanation for crustal folding. Du Toit concludes:

> 'It is nevertheless seriously doubted whether any of these other factors that are based upon the Earth's rotation, either singly or in combination, would be adequate to account for orogenesis or drift.'

With regard to convection currents, however, he states that:

> 'a system of magmatic streaming must develop whenever the viscosity of the material becomes sufficiently lowered...'

and cites the presence of the 'rapidly varying' Earth's magnetic field as evidence. Du Toit sums up as follows:

'The energy [for drift (author)] therefore must be derived from some internal source that is itself cyclic, and of such the only outstanding one is that provided by the Earth's radioactivity.'

The significance of radioactivity

In 1862, William Thomson, Baron Kelvin, assuming that the Earth was slowly cooling down from an originally molten state, had calculated from the present downward rate of temperature increase that the Earth must be between 20 Ma and 400 Ma old – subsequently revised to 20–40 Ma old. However, most geologists believed that the Earth must be very much older and that Kelvin's assumption of the conductivity of the rocks of the interior must be incorrect.

This problem is discussed by Arthur Holmes in his 1913 book *The age of the Earth*. Holmes showed that the key to the Earth's heat distribution, and consequently to its age, lay in radioactivity. Henri Becquerel had demonstrated the existence of radioactivity in 1896 by discovering that a sample of a uranium compound placed on top of a covered photographic plate emitted a mysterious radiation which was able to penetrate the black paper covering the plate, causing it to react as if it had been exposed to light. Marie Curie then examined all the then known elements and discovered that, in addition to uranium, thorium, polonium and radium were all radioactive.

The theory of disintegration of the radioactive elements was set out in 1902 by E. Rutherford and F. Soddy, who showed that the amount of radiation produced by a radioactive substance was proportional to the amount of the substance producing it and by nothing else. They also found that both potassium and rubidium were also radioactive, and in 1903, P. Curie and A. Laborde discovered that radium, which was known to be widespread within surface rocks, was able to maintain a temperature above that of its environment due to the kinetic energy of the radiation, and was therefore an important source of heat. This critical discovery at once invalidated all previous calculations of the strength of the Earth based on the cooling model.

Arthur Holmes

Holmes (1890–1965) pioneered the use of radioactivity in dating rocks, and published the first date using the uranium–lead method in 1911 shortly after graduating from Imperial College, London, where he continued to work on the subject. In 1913, he published *The Age of the Earth*, in which he estimated the age of the oldest rocks then known as 1600 million (1.6 Ga) years. By the 1950s this date had been revised upwards to 4500+/-100 Ma, approximately equivalent to the presently accepted date. In 1943 he was appointed Professor of Geology at Edinburgh University where he remained until retirement. His

Principles of Physical Geology was the standard textbook on the subject for many decades; the fourth edition, edited by P.McL.D. Duff, published in 1993, is still in use.

In his 1913 book on the age of the Earth, Holmes pointed out that, as the Earth had probably originated as a collection of solid particles ('plane-tesimals'), it was very unlikely that it had ever been completely molten. The internal heat would have initially arisen from the kinetic energy of the accumulating particles, and local fusion would result in certain constituents, such as the lighter and more viscous elements, moving upwards and the heavier metallic material moving downwards to form a nucleus. Thus a heavy metallic core would form, surrounded by a thick zone consisting largely of silicate compounds. However, Holmes believed that the subsequent thermal history of the Earth was controlled largely by its radioactivity.

The amounts of radium and thorium in various types of rock had been calculated by John Joly in 1909, from which he concluded that the total heat emission from the Earth amounted to 30×10^{-14} calories per second. This Holmes believed to be sufficient to raise the temperature of the Earth to $c.40,000°C$ over a period of $c.1000$ Ma, in which case the present geothermal gradient should be many times greater than it is, and could not be reconciled with what was known about the structure and temperature gradient of the crust now. It was more likely that the Earth is approximately in thermal equilibrium and is losing as much heat as it gains.

After considering the distribution of radioactivity within the Earth, Holmes concludes that, although the materials making up the interior must be relatively low in radium and thorium, if all the radioactive heat were supplied by an outer layer (i.e. the crust), this layer would only be 16 km thick, which would mean that its basal temperature would only reach 250°C – insufficient to produce melting. He therefore deduced that the radiogenic elements must also be distributed throughout the interior but would decrease in relative abundance with depth. This conclusion was consistent with the assumption of an upward concentration of the more acidic rocks such as granite, which are known to be more radioactive; in contrast, the more basic the rock, the poorer it is in radioactivity. He therefore concluded that these considerations were in accordance with an Earth model containing a thick ultrabasic mantle with a composition and radiogenic content similar to those of the stony meteorites (i.e. peridotitic) and a metallic core similar to the iron meteorites, and with even less radiogenic content. He estimated that almost all the radioactivity must be supplied by an outer layer with a thickness of around 30 miles (48 km), and that the basal temperature of this layer would be $c.750°C$, 'in accordance with the requirements of volcanic phenomena'.

The internal structure of the Earth

Our present model of the Earth's interior is derived largely from earthquake studies undertaken in the period between about 1900 and 1936. Until the later part of the nineteenth century, many geologists, including James Hutton, believed that the interior of the Earth was molten, and was overlain by a solid 'crust'. However, William Thomson, Baron Kelvin, had shown in 1864 that the Earth behaved in response to seismic waves with a rigidity comparable to that of solid steel! Had it been largely liquid, the effects of the gravitational attraction of the Sun and Moon would have affected both the continental and oceanic crust, to an extent that would greatly exceed the tidal effects of the oceans relative to the continents that we observe now.

The existence of a separate core was discovered by R.D. Oldham in 1906 by observing the existence of a shadow zone in seismic P-wave arrivals, and by 1935, H. Jeffreys and K.E. Bullen were able to produce seismic travel-time tables from which a model of the interior structure of the Earth could be deduced (Fig. 4.1). This was augmented in 1936 by Inge Lehmann's discovery of a discontinuity within the core from which she deduced that the inner core was solid. However, at the time that Joly and Holmes were discussing the heat distribution within the Earth, and the possibility of convection currents existing in a viscous medium, much less was known about Earth's internal structure.

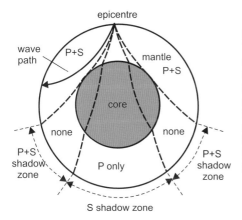

Figure 4.1 The interior of the Earth. The position of the core boundary is determined by the arrival pattern of P and S seismic waves. There are no P- or S-wave arrivals within a circular shadow zone on the opposite side of the Earth from the earthquake. As shear waves cannot be transmitted through a fluid, only P waves enter the outer core. Note that the crust on this scale is too thin to be shown.

The crust

The composition of the continental crust was known in some detail, since material from all depths of the crust is directly accessible at the surface somewhere. Continental rocks are extremely varied in composition, and include the whole range of igneous rocks together with all the different types of rock derived from them. The average composition has been estimated from considerations of the density distribution to be similar to a mixture of granite and basalt. The prevailing view in Holmes' day was that the upper part of the continental crust, called the 'sial' (from **si**lica plus **al**umina) was composed predominantly of lighter alumino-silicates in rocks such as granite, with an average thickness of around 10km. This layer rested on a more basic layer, the 'sima' (from **si**lica plus **ma**gnesia) where the denser magnesium-rich silicates predominated, composed of materials such as amphibolite, passing into eclogite (the high-density form of basalt) under appropriate conditions, and with an average thickness of 20–25 km. The oceanic crust, known to be of basaltic composition, was presumed to correspond to the sima.

The base of the crust was determined in 1909 by the Croatian seismologist Andreja Mohorovičić by observing that shallow-focus earthquakes produced two sets of earthquake waves, one of which followed a straight path to the surface while the other was refracted downwards by a higher-density layer (i.e. the mantle) (Fig. 4.2). The surface at which this change occurred was termed the Mohorovičić discontinuity or 'Moho' and lies at 5–10 km depth beneath the oceans and 20–90 km depth, with an average of *c.*35 km, beneath the continents. The depth was greatest beneath certain young mountain belts such as the Alps.

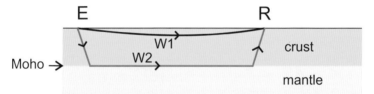

Figure 4.2 The base of the crust was determined by Mohorovičić when he discovered that earthquake waves originating at point E produced two sets of wave arrivals at point R: one set travelled through the crust, the other was refracted through the top of the higher-density mantle, arriving later at point R. The time difference is dependent on the depth of the crust.

The mantle

There was good evidence that the upper mantle is made of the ultrabasic rock peridotite. This material has the correct density, is known to melt to form basalt magma, and pieces of it are found within some basalts. It also forms the 'stony' meteorites – the most common type, and the probable source of much of the accreting material of the early Earth. Moreover, as we now know, in places where oceanic crust has been thrust onto continental margins, mantle material consisting of peridotite is found directly beneath the crustal rocks. At the time that Holmes and Joly were discussing the internal structure of the Earth, the term 'mantle' was not in general use; they referred to the region below the crust as the 'substratum'.

Peridotite is composed mainly of the minerals olivine and pyroxene, which are silicates of iron and magnesium. Other minerals probably also present in mantle peridotite include feldspar and metallic oxides. All the other elements found in the crust, including the heat-producing radioactive elements uranium, thorium and potassium, must also be present in small quantities, either within the main silicate minerals or in other compounds such as oxides, since the crust has presumably been formed over time from magmas derived from the mantle. There is a rapid change in seismic wave velocities within the mantle at between 400 km and 1000 km depth, which is interpreted as the effect of a downward change in the composition of the peridotite from olivine and pyroxene, characteristic of the uppermost mantle, to higher-density minerals such as spinel.

The core

The innermost, approximately spherical, region, the core, extends from a depth of 2900 kilometres to the centre of the Earth, at 6500 kilometres depth. The core is believed to consist mostly of metallic iron, with some lighter elements in addition, such as nickel. These constituents are thought to have become molten at an early stage in Earth's history and drained down towards the centre forming the core. In the outer core the metal is in a molten state but the inner core is solid, as discovered by Lehmann. Although the depth of the core boundary was known in Holmes' time, the internal structure was not.

Radiogenic heat and mantle convection

John Joly's contribution

John Joly (1857–1933) was an Irish physicist and professor of geology and mineralogy at Trinity College, Dublin. In 1899 he famously published one of the first estimates (80–100 Ma) of the age of the earth based on the amount of sodium in the ocean.

In a paper published in 1924, entitled *Radioactivity and the surface history of the Earth,* Joly pursues the consequences of radioactive heat for the properties of the Earth's interior. He suggests that, although the denser substratum beneath the granitic crust was now solid, as indicated from earthquake wave behaviour, there was no reason to believe that isostasy did not operate, which led him to the highly significant conclusion that:

> 'The highly heated elastic substratum may yield to crustal stresses of
> great magnitude maintained over long periods of time.'

He states that there could be no considerable escape of heat beneath the continents because they are themselves radioactive and their base must be maintained at a temperature closely approximating to the melting point of basalt. However, the situation was different beneath the oceans. Here heat is lost to cold ocean waters by conductivity, and thus an ocean floor of cooled basalt is formed which gradually grows in thickness with time as its base is maintained at the melting temperature of basalt. Joly supposed that the heat beneath both the continental and oceanic crust must be conserved, and that the substratum will gradually become fluid. Local convective movements would tend to facilitate the heat loss until finally a whole layer of the substratum is liquefied. At that point, he proposed that tidal effects from the Sun and Moon would operate to move the continents over the liquefied layer in order to equalize the heat loss between continental and oceanic areas. Rifts and flood basalt eruptions would form where the continents separated:

> 'Thus we find that the major events of a revolution (i.e. an orogenic cycle) find an explanation in the density changes which affect the substratum as the result of the alternate accumulation and dissipation of radioactive heat [and that] the source of periodicity [of orogenies (author)] lies in the facts that thermal discharge can only take place by convective movements and the co-operation of tidal effects and that of the extremely slow rate of supply that reversion to solidity becomes inevitable.'

Joly concludes as follows:

> 'the picture presented to us is not that of a slowly cooling world dying into quiescence. For if the Earth were, indeed, from the first a cooling planet and nothing more, as our forerunners supposed it to be, volcanicity and every manifestation of thermal or dynamic energy must dwindle age by age, till the only surface changes would be those due to solar heat and the tidal movements of the oceans. But we see no such decadence…We know that the continents and islands are still restless after scores of millions of years of thermal wastage.'

Arthur Holmes' contribution

Holmes returned to the problem of the temperature distribution in the Earth and its consequences in a ground-breaking paper in 1929 published in the Transactions of the Geological Society of Glasgow. He had now become familiar with Wegener's theory of continental drift and was one of very few contemporary geologists either in Britain or the USA to believe it.

In his paper, Holmes begins by stating that many hypotheses about continental drift ignore the Earth's internal heat, and that there was little recognition outside the United Kingdom of radioactivity as a heat source. He discusses various forces that had been proposed including tidal forces and pohlflucht, concluding that all of them were much too small to be effective. Even if they had an effect in the early stages of Earth history, some other and much stronger force must have operated subsequently. Thus he states:

> 'We may conclude that the dominant forces involved in crustal movements must arise within the Earth itself.... [But] ... in the early stages of the Earth history...the viscosity must have been very much lower and the crust must then have reached the stable position determined by a symmetrical distribution of the continents about the Equator. Since the continents are not so distributed at the present day, some other force must have operated in opposition to the pohlflucht to move them into the positions they now occupy...'

He raised a number of objections to the hypothesis of a cooling and contracting Earth, advocated by Harold Jeffreys, as outlined in the previous chapter: 1) the failure to explain the prevalence of plateau-basalt volcanicity (which implied considerable heat output throughout geological history); 2) the inability to produce sufficient contraction to explain the fold and thrust structure of the orogenic belts; 3) the fact that the distribution of orogenic episodes over time is episodic and thus different from that which would be deduced from the contraction hypothesis; 4) the improbability that compression could be dispersed through an outer layer 150 km thick and still produce superficial nappe structures such as those of the Alps; and 5) the failure to account for marine transgressions and regressions, and for 'geosynclines.' The final two objections were the most important: 6) the contraction hypothesis was incompatible with continental drift, in which Holmes was a firm believer; and 7) it also ruled out any heat output from the substratum despite the fact that there was no geochemical reason why the substratum should be so poor in radioactive elements that its radiogenic heat could be ignored.

In fact, the work of Holmes and Joly on radioactivity had demonstrated that the Earth must be much hotter than previous estimates based on the cooling Earth model. At higher temperatures, the strength of rocks diminishes, and the substratum could therefore be weak enough to flow. Evidence for this comes from the isostatic rebound of North America and Scandinavia since the withdrawal of their respective ice sheets around 20,000 years ago, during which time significant upward movements have occurred – up to 250 m in the case of Scandinavia (see Fig. 3.4). From this, Holmes calculates a value of viscosity for the mantle of between 5×10^{17} and 5×10^{23} – well below the limiting value of 10^{26} which Jeffreys had calculated would prevent convective circulation taking place. He concludes that the mantle is capable of transferring heat from the interior to the surface by slow flow in the solid state. In this he differs significantly from Joly, who had conceived of the upper part of the mantle being liquid.

Assuming that convection were possible, Holmes proposes that the effect of radioactivity would be to steepen the thermal gradient within the mantle, which would increase the rate of circulation until the point when the new heat could be carried off as fast as it was generated. However, the interior would only be able to discharge the excess heat if the crust could either be melted or broken through from below. He calculates, assuming the substratum has 1/700th of the heat content of plateau basalt, that:

> 'for the accumulated heat of 200 million years to escape through the sites of the oceans, one-third of the whole of the ocean floors (taken at 60 km thick) need to be heated up to 1000°C and replaced by magma which cooled to produce new ocean floors. This process implies some form of continental drift involving the sinking of old ocean floors in front of the advancing continents and the formation of new ocean floors behind them.'

Holmes proceeds to examine the mode of circulation within the substratum from general principles. He concludes that it would depend on 1) the varying longitudinal thickness of the substratum; 2) the varying thickness and radiogenic content of the crust; and 3) the Earth's rotation, which would tend to deflect upward currents westwards and downward currents eastwards, and deflect horizontal movements to the right in the northern hemisphere and the left in the southern (while admitting that these deflections were likely to be very slight).

Because the Earth is not a perfect sphere, but is slightly flattened at the poles due to the centrifugal force caused by its rotation, the thickness of the substratum at the Equator is greater than that at the poles, and because its radiogenic content should be independent of latitude, the equatorial temperature gradient should be steeper than that at the poles. This led Holmes to his proposed model of convection, shown in Figure 4.3. He points out that in a layer of viscous liquid, heated uniformly, and bounded by rigid conducting surfaces above and

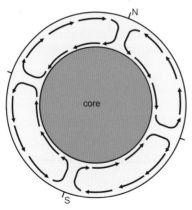

Figure 4.3 Holmes' model of convection in the Earth's mantle. After Holmes, 1929.

below, no convective circulation can occur until a critical temperature gradient is reached, which will depend on the compressibility, conductivity and viscosity of the liquid. He concludes that when this critical point is reached, the layer will become unstable and convective circulation will begin, producing a system of cells as shown in Figure 4.3. He proposes that as the ascending equatorial currents approached the base of the crust, they would divide into two radial branches: one northwards and the other southwards. As the opposing northwards currents met around the polar region, they would turn downwards, and the same would happen to the southward-moving currents. In an ideal initial state, a system of two cells with circular axes would be created, one in the northern hemisphere and the other in the southern. The upper parts of such a system of convection currents could, he thought, carry continents across the Earth's surface. In the case of a continent originally lying across the Equator, the effect would be to split it and drag the pieces apart, leaving a depressed 'geosynclinal' belt or a new ocean between. As well as providing the missing mechanism for continental drift, this idea was a major contributory factor in the development of the plate-tectonic theory, discussed in chapter 6.

Holmes then considers the effects of the higher radiogenic content of the continental crust, which he believed would create a secondary system of convective currents beneath the continents in order to carry away the excess heat created there towards the continental margins, as shown in Figure 4.4. Where

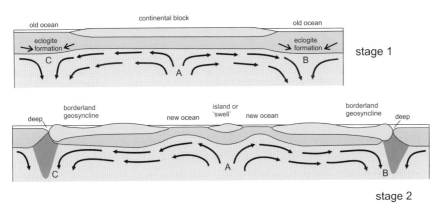

Figure 4.4 Holmes' model of sub-crustal circulation. Upper crustal layer (*sial*), yellow; 'intermediate layer' (gabbro etc.), green; 'substratum', purple. Note that this is the terminology of Holmes' day and does not correspond to that of today. In stage 1, convection currents rise beneath a continent and descend at its margins. In stage 2, the continent has been 'distended' on each side of A, leaving an island or 'swell' in the 'dead zone' above A. Above B and C, eclogite formation results from the crystallization of the material of the intermediate layer, and the oceanic deeps (i.e. the ocean trenches) are produced. The front part of the sial is thickened and a 'borderland' geosyncline is formed behind it. One effect of the heat transport from A to B or from A to C is the development in each case of a geosyncline. After Holmes, 1929.

the ascending currents turn sideways, the opposed flowage in the crust would create a stretched region, or 'disruptive basin', which could ultimately become a new ocean, and discharge a great deal of excess heat.

Where currents meet and turn downwards, especially at continental margins, Holmes suggests that the crust above the downturns would experience strong compression, resulting in the thickening of the amphibolite layer, which, under the combined effects of high temperature and pressure, would be expected to lead to the production of eclogite. This change involves an increase in density from 2.9–3.0 to at least 3.4, and would lead to marked subsidence. He concludes:

> 'Such foundering would effectually speed up the downward current for two reasons: the greater density of the sinking blocks, and the cooling of the substratum material in their vicinity.'

This process also offered an explanation for the existence of the deep ocean trenches that accompany the active orogenic arcs around the Pacific margins, and Holmes notes that the deep earthquakes off the coast of Japan could be connected with the foundering of the denser blocks. This insight was finally to be justified by the discovery in the 1960s of the Benioff zone and its significance for the process of subduction, discussed in chapter 6, which led directly to the plate-tectonic theory.

Other consequences of the convection process would be experienced in the continental crust, which Holmes believed would be thickened near the continental margins, and thus create mountain roots that would begin to melt, producing igneous activity of the 'Circum-Pacific type' exemplified by the basalt–andesite–rhyolite suite of volcanicity. One of the criticisms of Wegener's model had been that if the material of the oceanic crust was weaker than that of the continents (as his model required), then mountain building could not take place at the continental margins, as it does along the western coasts of the Americas. Holmes' proposal removed this difficulty by placing the origin of the compressional force at a deeper level, i.e. within the 'substratum'.

As the region above the downgoing currents subsides and fills with sediments, the conditions are created that correspond to the classic model of the geosyncline: a thick marine sediment pile is combined with volcanicity derived from the melting of the hot substratum and also, ultimately, from the lower layers of the sediment fill. The compression that was likely to accompany this process could explain the folding that was known to accompany geosyncline formation but for which 'hitherto no explanation has been offered'.

Holmes also attempts to calculate the likely velocity of the sub-continental currents and comes up with the surprisingly familiar figure of 5 cm per annum – close to an average velocity for the present plates. He then goes on to visualize how the system might evolve with time, suggesting that the zone of subsidence would tend to migrate outwards from the continental margin and eventually disappear, causing the current to slow down and to be replaced by an opposing current. At the same time, as the new ocean floor cools, the ascending current there will fade for lack of 'thermal sustenance'. The consequence would be that 'gradually an entirely new convective system will evolve, tending to close up the continents again'. He cites the approach of Africa and Europe across the site of Tethys as an example of this.

Vening Meinesz's views on mantle convection

After the publication of Holmes' 1929 paper, attempts were made to investigate mantle convection from a theoretical standpoint, but no significant advance could take place until further progress had been made towards an understanding of the internal structure of the mantle. Nevertheless, an interesting contribution to this subject was made in 1962 by F.A. Vening Meinesz, who undertook a detailed theoretical analysis of the possible geometry and mode of operation of the convection process.

Vening Meinesz

Felix Andries Vening Meinesz (1887–1966) was a distinguished Dutch geophysicist who, having designed an improved type of gravimeter, undertook the first detailed gravity survey of the Netherlands in the 1920s followed by a marine gravity survey, along with other Dutch earth scientists, of the then Netherlands East Indies. There, he made the significant discovery that the deep ocean trenches were associated with a negative gravity anomaly (i.e. they were not isostatically compensated), implying that a downwards force was acting to disturb gravitational equilibrium. He was appointed professor at Delft University of Technology in the 1930s, and after the Second World War became the director of the Royal Netherlands Meteorological Institute.

In the 1962 volume *Continental Drift*, edited by S.K. Runcorn, Vening Meinesz first reviews the arguments in favour of convection currents. He assumes that the continued production of basalt by selective fusion of peridotite would eventually exhaust the upper mantle source layer, unless it were renewed by a supply of fresh hot material from below, taking that as convincing evidence for the existence of such currents. He also follows Holmes in rejecting the contraction theory as an explanation for geosynclines and orogenesis. He makes an important distinction between 'oceanic' geosynclines and those situated on continental crust:

> 'The fact that the oceanic crust consists nearly entirely of olivine and therefore has nearly the same mean density as the mantle, makes it

possible that in each oceanic geosyncline the down-buckling crust penetrates deeper into the mantle than is the case for the continental geosyncline where the mean density is so much less.'

A convection cell model

In his 1962 paper, Vening Meinesz summarizes the physical properties of the mantle in terms of their likely role in the convection process, pointing out the importance of the transition layer (between c.500 km and c.900 km depth) within the mantle where the olivine–pyroxene assemblage must give way to higher-density equivalents. He concludes that this layer would be a much stronger source of instability than the effect of cooling mantle at the surface, and suggests that without this layer, convection currents may not have existed.

He examines how this process would work, supposing that if a current is started by some secondary phenomenon, subcrustal cooled material would enter the subsiding column and higher-temperature material would enter the rising column to replace cooled material there. As the mean densities of the two columns begin to differ, the subsiding column becomes heavier and the rising one lighter, favouring continuation of the flow, which would tend to increase in velocity until the current has made about a quarter turn. At that point, the major part of the lower-temperature, higher-density material is located within the subsiding column and the higher-temperature, lower-density material in the rising column. From there, the velocity would gradually decrease until, after the second quarter turn, all the low-temperature material is in the lower mantle layer and all the higher-temperature material in the upper layer. The current would then stop, the system having reached dynamic stability.

The thermal conditions, however, would now be unstable, since the lowest-temperature material at the base of the mantle is now in contact with the high-temperature core, while the hottest part of the mantle is in contact with the cooler crust. In both parts of the system, strong heat conduction must occur which, although very slow, will gradually change the situation, leading to smaller convection cells, which themselves will act to eventually restore the conditions that led to the half-turn convection – that is, to a mantle with lower temperatures in the upper part and higher in the lower part. Again, some secondary phenomenon must then act to bring about a second cycle of half-turn convection. Because the known cycles of strong tectonic activity generally last for about 100 Ma and are separated by periods of about 300 Ma, Vening Meinesz deduces that the periods of more active convective activity would last for c.100 Ma and the periods of relative quiescence for c.300 Ma.

Turning to the question of the size and shape of the convection cells, Vening Meinesz assumes that each individual cell rotates as a solid, and that

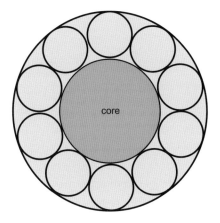

Figure 4.5 Vening Meinesz's 10-cell convection system. Each cell is envisaged as part of a circular tube-like structure, which would partially encircle the core, parallel to a great circle. Adjoining cells would rotate in opposite directions. After Vening Meinesz, 1962.

only the outer layers of the cell behave as a viscous fluid. In contrast with Holmes' model, he concludes that the cell diameter would be equal to the mantle thickness, and proposes a pattern of ten rotating cells, as shown in Figure 4.5. In seeking evidence to corroborate this pattern, he examines the variation of the Earth's shape.

The shape of the Earth

The Earth is not perfectly spherical, but is flattened at the poles due to the centrifugal effect of its rotation. Moreover, the surface shape is obviously not uniformly smooth, but exhibits large-scale topographic variations that are not randomly distributed, the most obvious being the difference between the deep oceanic basins and the continental platforms. Smaller-scale topographic anomalies are represented by the mountain ranges, ocean ridges and deep ocean trenches. This topographic variation can be expressed as a series of spherical harmonic terms with various frequencies, like sets of waves with different amplitudes and wavelengths. It appears that there is a periodicity in the height distribution, which corresponds to peaks of elevation with frequencies (n) of 1, 3, 4 and 5. These values are independent of the effect of the polar flattening.

Vening Meinesz thought that this pattern must reflect the pattern of mantle convection, and explains this as follows. In the early history of the Earth, when the whole interior behaved as a viscous fluid, the continental crust would have been concentrated in a single mass by a single convective cell (corresponding to a pattern of $n=1$) with its uprising column beneath the single universal ocean and down-turning current beneath the single continent. As the core began to grow, and the base of the mantle moved upwards, the convective system would have changed to a whole-mantle system with smaller cells with values of $n=3$ and 4, causing the supercontinent to split into separate continents and new

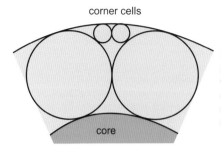

corner cells

core

Figure 4.6 Vening Meinesz suggests that small secondary 'corner cells' would develop where the primary circulating currents converged or diverged, rotating in the opposite direction to the adjoining primary cell. After Vening Meinesz, 1962.

oceans to form. At a later period, when the mantle had attained its present size and solidified, the convective cells would again redistribute themselves into a pattern determined by the mantle width, corresponding to a frequency of $n=5$, equivalent to the ten-cell pattern of Figure 4.5. The higher-frequency peaks in the distribution were visualized to correspond to tectonic features such as the deep basins in island arc regions, caused by small corner cells as shown in Figure 4.6.

Later work, influenced by much more information about the internal properties of the mantle, has shown that the pattern of mantle convection is much more complex than Vening Meinesz had envisaged; however, his attempt to link the convection pattern with geological and topographic features was an important step in understanding this fundamental process.

The influence of plate tectonics

The introduction of the plate-tectonic theory in the late 1960s transformed ideas on how convection would work. It became clear that the behaviour of a strong, cool lithosphere shell (see chapter 6) would partly control the convection process and that, contrary to what Holmes had thought, the force exerted by a weak viscous layer of moving mantle material was much too small to shift the continents. Instead, the gravitational effects of sinking lithosphere slabs and raised warm ocean ridges were the main source of continental movements. Further progress in understanding how convection works would depend on future discoveries involving more sophisticated geophysical and geochemical methods and will be the subject of chapter 10.

Nevertheless, Holmes was the first to offer a detailed and convincing defence of the continental drift theory, based on a plausible mechanism. It is remarkable that his contribution was not universally recognized until over thirty years later, when geophysicists finally abandoned their attachment to the strong earth model, provided palaeomagnetic evidence in support of drift, and considered convection as a viable process.

Postscript

The discoveries of the early researchers in radioactivity were crucial to the understanding of the Earth and were recognized by the award of the Nobel Prize: jointly to Marie Curie, her husband Pierre, and Henri Becquerel in 1903, to Ernest Rutherford in 1908, for their work on radioactivity, and to Marie Curie again (for Chemistry), in 1911. Arthur Holmes became a Fellow of the Royal Society in 1942.

5

Deformation ellipsoid to ductile shear zone

Derek Flinn (1922–2012)

A paper entitled *On folding during three-dimensional progressive deformation* was read at the Geological Society of London in 1962 by Derek Flinn (in these days, papers accepted for publication by the Society had to be read first at a meeting of the Society, where they were subjected to criticism). The first to comment on Flinn's paper was John Ramsay (subsequently to become Professor of Geology at Imperial College, London, and thereafter at ETH, Zurich). Ramsay prefaced his detailed response by stating that Flinn's paper was 'perhaps one of the most significant contributions made to theoretical structural geology for many years'. Ramsay himself was, and is, one of the outstanding structural geologists of his generation, so his praise was all the more cogent. According to Jane Plant, Flinn's obituarist, it was said at the time that Ramsay and Flinn were the only two people who could understand each other's papers!

Perhaps because of the excessively mathematical nature of Flinn's treatment of strain, little use was made of his methods until Juan Watterson employed them to describe the deformation of the Precambrian gneisses of West Greenland. It is largely the work of these two Liverpool University geologists that has revolutionized the methods by which the highly strained rocks of the major shear zones have subsequently been described and analysed.

After a period of war service in the Royal Marines, Flinn studied geology at Imperial College, London, graduating in 1950 with a first-class honours degree, after which he went on to obtain a PhD there. The late Professor Janet Watson, herself a distinguished geologist, is said to have described Flinn as one of the brightest students of his generation. In 1953, Flinn was appointed as lecturer at Liverpool University, where he spent the remainder of his career, ultimately being awarded a personal chair in 1975.

Most of Flinn's work centred on the geology of the Shetland Islands, much of which he mapped himself in detail. However, it was during an exchange Fellowship at the Institute of Geology and Mineral Deposits in Moscow in 1960 that he wrote his key paper on the deformation ellipsoid. His application of the ellipsoid to the description of strain is said to have stemmed from his experience in teaching crystallography.

Flinn's theory of progressive homogeneous deformation

In his 1962 paper, Flinn observed that structural geologists generally make assumptions about the nature of strain, dealing with it as if it was essentially two-dimensional. He was critical of the then-current system of describing deformed structures according to the three-axis a-b-c system (attributed to Swiss geologist Bruno Sander) where the b-axis equates to a fold axis and the a-axis is the direction of compression, or transport. In describing progressive deformation (i.e. strain), he showed that it was necessary to allow for the planes and lines of the strained object both to rotate and to change dimensions during the deformation process.

Homogeneous deformation

Flinn's model is applicable only to 'homogeneous' deformation (i.e. homogeneous strain), during which a material deforms by means of a process usually referred to as 'flow' or, more strictly, as 'viscous' behaviour, where the amount of strain is directly proportional to the size of the applied stress and the length of time during which the stress is applied. This type of deformation is permanent, and contrasts with 'elastic' deformation, which is temporary.

Structural geologists usually apply the term 'flow' in rocks to behaviour similar to that exhibited by the movement of a liquid of low viscosity but on a vastly different timescale. A useful analogy for homogeneous strain is with the flow of a river constrained between straight and parallel banks, where the flow velocity varies smoothly and regularly from its fastest at the surface, in the centre of the river, towards its slowest at the margins or at the bottom. It is this type of flow that Flinn describes, although it should be recognized that more irregular types of flow (i.e. 'heterogeneous' deformation) are actually more common in nature.

Homogeneous deformation is only found near the Earth's surface in rocks with exceptionally low viscosity, such as glacier ice and salt, and even then only in certain parts of the rock mass. Viscous flow in glacier ice is well illustrated in Figure 5.1 where the differently coloured bands of ice have been distorted into 'flow folds'. Elsewhere in upper-crustal rocks, deformation is typically controlled either by faulting or by the buckling of the stronger layers in a sequence of strata, leading to a highly heterogeneous strain distribution. However, in the warmer and deeper levels of the crust within the orogenic belts, and especially in relatively homogeneous masses of igneous or gneissose rocks, viscosity contrasts between the various components are much less important, and the conditions for ideally homogeneous deformation can exist. Even in regions of

Figure 5.1 Folding in glacier ice in the Lewis glacier, Alaska. Note that the folds have a 'similar' form, i.e. each layer is the same shape as the next one. This is a characteristic of flow folds produced by homogeneous deformation. © Jim Wark.

heterogeneous deformation, the application of Flinn's model of progressive homogeneous deformation can provide a useful approximation to the overall strain pattern. His analysis has proved to be particularly useful in the study of the highly strained ductile shear zones typical of regions of Precambrian granites and gneisses, as described below.

The characteristics of ideal homogeneous strain are that the movement of particles within the strained body is regular and continuous during the process of deformation, varying smoothly from one part of the strained body to another. Consequently, lines existing before deformation remain straight, and planes remain planar. Thus folds cannot be generated during this type of deformation, although folds already present will be deformed into a different shape. The scale of observation, however, is critical, since strain may appear continuous at the scale of the outcrop or hand specimen yet be discontinuous under the microscope!

The deformation ellipsoid

More usually known as the 'strain ellipsoid', this device is a method of portraying strain in three dimensions, assuming the unstrained body to be represented by a sphere (Fig. 5.2). Flinn credits the Norwegian geologist Hans Ramberg for first indicating how geological structures could be fitted into a model of three-dimensional strain using the deformation ellipsoid, but it was Flinn who developed the method into a comprehensive and usable theory.

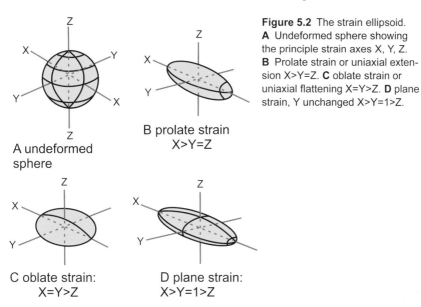

Figure 5.2 The strain ellipsoid.
A Undeformed sphere showing the principle strain axes X, Y, Z.
B Prolate strain or uniaxial extension X>Y=Z. **C** oblate strain or uniaxial flattening X=Y>Z. **D** plane strain, Y unchanged X>Y=1>Z.

The ellipsoid has three mutually perpendicular axes, e.g. X, Y and Z, where X>/=Y>/=Z.* Certain types of strain can thus be represented as either 'prolate' ellipsoids (i.e. shaped like a cigar, or a rugby-ball), 'oblate' ellipsoids (i.e. shaped like a pancake, or a curling stone). In prolate ellipsoids, X>Y=Z, and the strain is constrictional, in the sense that both Y and Z axes have been shortened (Fig. 5.2B); in oblate ellipsoids, X=Y>Z, the strain is flattening, and both X and Y axes have been extended (Fig. 5.2C). A special type of strain is known as 'plane strain', where X>Y>Z, but the length of Y has remained unchanged (Figure 5.2D).

In the case of progressive homogeneous strain, an original sphere changes shape progressively through a series of ellipsoids that vary smoothly and continuously in shape until the deformation ceases. During this process, linear structures will, in general, move away from the Z direction towards the X direction, whereas the poles (perpendiculars) to planes will move away from X towards Z. Figure 5.3 shows this process in two dimensions. However where planes or poles initially lie within the symmetry planes of the ellipsoid, they will remain within these symmetry planes.

Deformation paths

Flinn derived mathematical formulae to describe the shapes of the various ellipsoids produced during progressive deformation, and the equations for the

*Note that the author has followed John Ramsay and subsequent authors in reversing Flinn's nomenclature for the fabric axes. Flinn uses the system Z>Y>X, where Z is the greatest and X the least; this is now usually shown as X>Y>Z, where X is the greatest and Z the least.

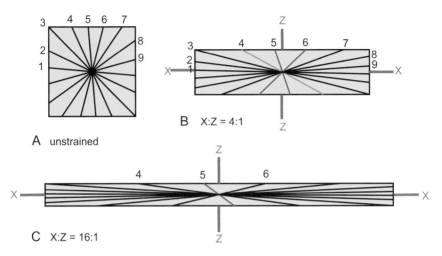

Figure 5.3 Effects of progressive homogeneous strain in two dimensions. **A** Unstrained square containing a set of lines distributed evenly at 20° intervals. **B** Distribution after strain of X:Z=4:1. **C** Distribution after a strain of X:Z=16:1. Note that at 4:1, the lines begin to concentrate around X, most are elongated, but green lines 5 and 6 are shortened and orange line 4 remains the same length; at 16:1, all the black lines are elongated and concentrated within 6° of X except for green line 5, which is shortened.

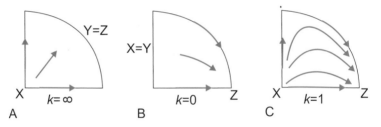

Figure 5.4 Movement paths for poles to planes for progressive strain: **A** prolate strains k=∞; **B** oblate strains, k=0; **C** plane strain, k=1. For lines, the movement paths are reversed. After Flinn, 1962.

movement paths of planes or lines during the deformation. If the axial ratios of an ellipsoid are: $a=X/Y$ and $b=Y/Z$, then the shape can be described by the value: $k=(a-1) / (b-1)$. The special cases of $k=0$ (oblate uniaxial ellipsoids), $k=1$ (ellipsoids where Y is unchanged) and $k=\infty$ (prolate uniaxial ellipsoids) are shown in Figure 5.4. The movement paths taken by planes or lines during progressive deformation can be shown graphically. Poles to planes always move away from X and towards Z, whereas lines move away from Z towards X. This happens after relatively low strains: thus at $b=10$ and $k=0$, half the poles will have become concentrated within 10° of Z; for $k=\infty$, poles are concentrated near YZ but to a lesser degree: at $a=10$, only 7% are concentrated there.

Flinn showed that strains could be conveniently displayed in two dimensions by using what has become known as the Flinn diagram where the

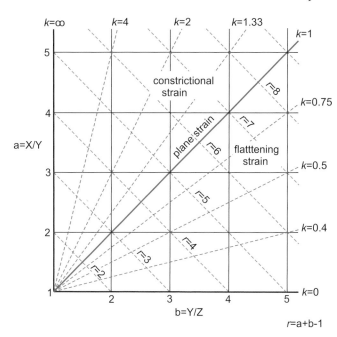

Figure 5.5 Flinn diagram. All types of strain ellipsoid and strain magnitude are shown in two dimensions by plotting a=X/Y against b=Y/Z. Types of ellipsoid given by values of $k=(a-1)/(b-1)$ from $k=0$ to $k=\infty$ shown by dashed green lines and values of $r=a+b-1$ giving magnitude shown by red dashed lines. The origin represents the undeformed sphere with radius=1. After Flinn, 1962.

measured values of *a* are plotted against those of *b* (Fig. 5.5). The diagram is divided into fields of constrictional and flattening strain separated by the line representing plane strain. If the stress state does not vary throughout the course of the deformation, the axial ratios *a* and *b* do not change and the deformation path is represented in the Flinn plot by a straight line. A set of such lines representing the varying values of *k* radiate from the origin. The magnitude of the strain is represented by a set of lines showing increasing values of $r=a+b-1$. Thus any strained object, or set of objects, from which values of axial ratios *a* and *b* can be obtained (such as ooliths or pebbles) can be used to estimate the shape of the strain ellipsoid and the magnitude of the strain that characterize the deformation.

Generation of folds

Folds cannot be generated during homogeneous strain, so any folds found within the deformed body must have been present before the homogeneous phase of deformation commenced. What happens to such folds can be shown by representing the folds as bundles of planes intersecting on the fold axis. These planes can move in such a way that the initial fold may open up during progressive strain, or it may become tighter.

One of the more important observations made by Flinn was that traditional methods of discovering the unstrained state of a folded body by unrolling it about its fold axis was valid only for the very limited case of plane strain.

For all other types of strain, the fold axis itself will have changed its length and the reconstruction will be incorrect.

Flinn's conclusions

1) *Preferred orientation.* This term refers to a pattern where lines or planes form a cluster in a particular zone of the strain ellipsoid rather than forming a random distribution. Flinn describes this state as follows:

> 'in a homogeneous mass of rock containing randomly oriented planes, flakes or rods…[these] develop a preferred orientation [which] changes in character as the strain changes and the two have the same symmetry.'

Thus, for example, if it could be assumed that a set of veins within a homogeneous mass of granite was randomly oriented before deformation, even if initially buckled or subjected to 'boudinage' (i.e. extended), each vein would rotate into its appropriate attitude to form a preferred orientation.

2) *Folding.* During progressive homogeneous deformation:

> 'Straight lines and planes rotate towards the longest axis of the deformation ellipsoid but remain straight and planar. Parallel lines rotate but remain parallel. A curved surface, however, rotates bodily and either becomes more or less curved depending on its attitude…so that the fold axis and the axial plane approach the longest axis of the deformation ellipsoid and the layer thickens or thins and folds up further or unfolds about the same axis. If the folded layer had parallel boundaries before the deformation, the boundaries must be parallel afterwards and thus the folding is of similar type.'

Flinn showed that, because initially planar surfaces must remain planar, folds cannot be generated during progressive homogeneous deformation, so that all folds found as a result must have been initiated by some other mechanism, such as 'buckling', during the preliminary stages of the deformation, when the viscosity contrasts between layers were sufficiently large to enable inter-layer slip to take place. In the buckling process, parallel layers are subjected to compressive stress acting parallel or subparallel to the layers, and these remain parallel after being folded. He notes that folds generated by buckling and then subjected to homogeneous strain will exhibit a mixture of parallel geometry inherited from the buckling stage and similar geometry produced by the superimposed homogeneous strain. In 'similar' folds, the folded layers are no longer parallel after folding but successive fold shapes are similar. Such folds may become more closed, or may open up about their original fold axes.

3) *The deformation ellipsoid and the tectonic axial cross.* It follows from Flinn's discussion of how planes and lines behave during progressive homogeneous strain that neither fold axes nor axial planes bear any special relationship to the axes of the deformation ellipsoid and are therefore of no value in determining the direction of movement or flow in deformed rocks. Consequently the traditional method of determining movement direction by the tectonic axial cross (a-b-c), where b is the fold axis, and the movement direction a is perpendicular to b, is equally valueless, except under the special circumstances of upper-crustal rocks where the deformation is dominated by the folding of highly competent, strong layers. Accordingly:

'The determination of the shape and orientation of the deformation ellipsoid must be made from the structures formed during that deformation.'

Such structures include: a) folds (compressed layers) and boudinage (extended layers) – only if these were parallel to the symmetry axes of the ellipsoid before deformation; b) the preferred orientation of minerals (such as micas) that have grown within the rock during the deformation and that would be expected to form parallel to the Z–Y plane (i.e. the plane of flattening); c) objects such as oolites, pebbles, pigment spots, grain shapes, and certain fossils, that can give a direct indication of the shape and orientation of the ellipsoid, especially if such objects are likely to have been spherical or nearly spherical before deformation.

4) *Flow folding.* Flinn demonstrates that folds with similar geometry develop through a process of flow parallel to the axial plane of the fold, and also parallel to any axial-plane cleavage that is formed as a result of flattening in the X–Y plane of the ellipsoid. This proves that the alternative explanation, advocated previously by many structural geologists, that similar folds are formed as the result of shear stress acting obliquely to the axial plane, is incorrect.

5) *Complexity.* It might be thought that, given the large number of possible variations of orientation and shape presented by the deformation ellipsoid, structures found in nature should be much more complex than they actually seem to be. Flinn suggests that this may be because the layering and fold axes of the deforming strata in many areas had occupied special positions in the deformation ellipsoid at the start of the homogeneous deformation: for example, if the layers were initially parallel to Y–Z and the fold axes to Y, and that once folding had commenced under conditions of heterogeneous deformation (e.g. by buckle folding), the resulting structure was strong enough to resist attempts to impose a different regime upon it, even although the regional stress system may have required it.

Applications to the West Greenland gneisses

Two years after Flinn's revolutionary paper was published, his young Liverpool colleague, Juan Watterson, began to apply the new methods to the Precambrian gneiss complex of Southwest Greenland. Working under the auspices of the Geological Survey of Greenland, Watterson took part in the Survey mapping programme in the Frederikshåb area, part of which resulted in his paper entitled *Homogeneous deformation of the gneisses of Vesterland, Southwest Greenland*, published in 1968. This publication clearly demonstrates the value of Flinn's methods in understanding the true nature of these types of Precambrian gneiss terrains and in unravelling their complex structural history.

The gneiss complex of Vesterland

The area described by Watterson is about $100\,km^2$ in area and comprises the large island of Vesterland together with several adjacent smaller ones. It is situated south of the town of Frederikshåb within what is now known to be the Archaean block of central West Greenland, between the Palaeoproterozoic belts of the Ketilidian to the south and the Nagssugtoqidian to the north. At the time of Watterson's work, all that was known about the regional setting of the area was that it was pre-$1700\,Ma$ in age.

The rocks described by Watterson consist mainly of banded or foliated gneisses of granodioritic composition, with subsidiary masses of basic and ultra-basic gneiss. The gneiss is said to be 'migmatitic', which is a way of describing the presence of numerous veins, bands or lenses of granitic or quartzo-feldspathic material that may be either concordant or discordant to the dominant planar structure. The major structure is that of a large open fold with a steep, highly sheared and mylonitized eastern limb, designated F3 in the author's structural sequence (Fig. 5.6). The many excellent photographs provide informative illustrations of the rock types and structures.

The gneisses display a penetrative planar fabric with a strongly developed elongation lineation. The fabric consists mainly of the oriented ellipsoidal shapes of the quartzo-feldspathic lenses (i.e. 'augen') in the gneisses and the elongated shapes of mafic bodies. In terms of the strain ellipsoid, the X fabric axis is generally much longer than the Y axis, which is closer to the Z axis (i.e. $k > 1$) but the shapes are quite variable. The plane containing the X–Y axes corresponds to the axial plane foliation of both F1 and F2 folds and nearly everywhere parallels the gneiss banding. The dominant elongation lineation is also parallel to the F1 and F2 fold axes. The composite planar structure, which includes both the gneissose banding and the axial-plane foliation of the F1 and F2 folds, is folded around the major open F3 fold. However, where the earlier fabrics are affected by strong F3 deformation, the elongate aggregates that define the earlier structure have become parallel to the F3 stretching direction. Thus Watterson

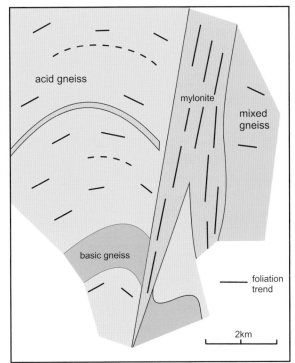

Figure 5.6 Very simplified sketch map showing the main features of the structure of Vesterland. Note that the mylonite belt corresponds to the attenuated middle limb of an asymmetric double fold, shown clearly by the basic gneiss band. The margins of the mylonite belt are gradational. Based on Watterson, 1968.

concludes that the total strain pattern, including the shape elements, must be regarded as the cumulative effect of successive strains, rather than as the product of a single episode.

The different rock types respond differently to the high strain: in the acid granitic gneisses, both the post-F1 and post-F2 deformations have been sufficient to completely re-orient the F1 and F2 axial planes into parallelism with the XY plane and the F1 and F2 axes into parallelism with the X direction. In the mixed gneisses, post-F2 deformation has re-oriented the previous axes but only incompletely re-oriented the axial planes. In the basic gneisses neither the axes nor the axial planes have been re-oriented, indicating that the existence of the more competent basic material within the gneiss complex has reduced the effective intensity of the deformation.

The gneisses are cut by a large number of discordant veins of pegmatite or aplite thought to be broadly contemporaneous with the F1–F2 deformations. These have a common intersection parallel to the fabric X-axis. On surfaces parallel to X, the veins appear to be parallel and are unfolded, but on surfaces perpendicular to X, the veins have a random orientation and some are folded. This is explained as a result of the homogeneous deformation, which has re-oriented all the planes in such a way that their strike is parallel with the stretching direction X, but has made little change to their orientation in the YZ plane.

The strain ellipsoid

As a result of the high degree of strain, in many places the fabric elements of the earlier phases of deformation have been re-oriented into parallelism with later elements in such a way as to be indistinguishable from them. Thus, the F1 axial-plane foliation and the F2 axial plane foliation are now generally parallel to the original banding. These planar structures have locally been re-orientated as a result of the F3 deformation but the linear elements are generally sub-parallel to the F3 stretching direction over most of the area.

The F3 folds are considered to be genetically related to the mylonite belt: the attenuated middle limbs of the small-scale asymmetric F3 folds are mylonitized to mirror the way that the mylonite belt represents the attenuated middle limb of the major F3 fold. The foliation in the mylonites and the axial planes of the minor F3 folds are sub-parallel to each other and also to the axial plane of the major fold. Watterson concludes that the mylonitic banding probably represents the XY plane of the deformation ellipsoid and that the stretching direction in the most highly deformed rocks represents the X direction and thus can be considered as the kinematic movement direction.

The strain ellipsoid is considered to be a combination of the ellipsoids representing all the deformations rather than simply the F3 strain. Applying Flinn's method of strain analysis to the Vesterland area, Watterson calculates that the essential characteristics of the deformation can be portrayed by a strain ellipsoid with values of $k=17$ and $r=37$, where k is a measure of the shape and r (where $r=a + b - 1$) is a measure of magnitude (see Fig. 5.5). This corresponds to an ellipsoid shape with axial ratios 1:3:100 – i.e. a highly prolate or constrictional strain (Fig. 5.7).

Conclusions

Watterson summarizes the results of his meticulous and well-argued account as follows:

> '…an area of common-place gneisses has i) been deformed in a manner which can be represented as a three-dimensional finite homogeneous strain, and ii) that the minimum amount of deformation, r, is in the region of 35.'

He suggests that this type of deformation is probably typical of basement rocks of the Precambrian shield regions:

> 'One of the characteristic features of gneissose rocks is the prevalence of a banded or streaky structure which is frequently shown on a variety of scales from 1 mm to 1 m or more, often within the same rock … [and] … This characteristic can be seen as the inevitable result of intense deformation of the type demonstrated in Vesterland; with

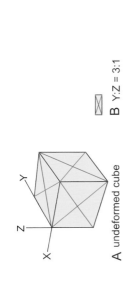

A undeformed cube

B Y:Z = 3:1

C X:Y = 100:3

D X:Z = 100:1

Figure 5.7 Effects of a deformation with strain ratios 1:3:100 (Watterson's estimate of the total strain in the Vesterland rocks) on a cube with red and green diagonals on each face (**A**). Note that discordances can be clearly seen on the Y:Z face (**B**), less clearly on the X:Y face (**C**), but are virtually indistinguishable on the X:Z face (**D**). Thus on rock surfaces parallel or near-parallel to the XZ plane, all lines and traces of planes will appear to be concordant.

deformation in which r>20 any original inhomogeneity of whatever shape or pattern is likely to give rise to a banding or streakiness; any previous planar structure may be re-oriented into parallelism with the new banding. The result is…the production of deceptively simple regularly banded gneisses from heterogeneous and perhaps intricately folded schists and migmatites, purely as a result of deformation.'

This statement revolutionized the way that geologists working in areas of Precambrian basement rocks viewed the origin of such rocks. Not very many years previously, a great debate had raged between those who believed that many (if not all) banded gneisses and migmatites had been formed as a result of a process of 'granitization' where metasomatic processes transformed sedimentary rocks of the right composition (e.g. greywackes) into granitic gneiss, or even homogeneous granite. Ascribing their layered structure to deformation removed one of the main arguments in favour of a sedimentary origin for such gneisses. Another important conclusion is that intense deformation is likely to produce a deceptively simple picture of the stratigraphic relationships, concealing structures such as unconformities and discordant dykes by reducing or removing their original discordance.

Not long after Watterson had begun to apply Flinn's methods to the West Greenland rocks, John Ramsay incorporated them into his textbook on structural geology, published in 1967, after which they soon became part of the structural geologist's toolkit for investigating highly strained rocks.

Looking at strain in terms of a homogeneous deformation, even although this may have been an approximation, made geologists aware of how the various structural elements of a particular deformation: folds, boudinage, and fabric (e.g. slaty cleavage, etc.) could be integrated into an overall picture (Fig. 5.8).

Applications to ductile shear zones

It was quickly realized by structural geologists that the reason why such high strains were encountered in Precambrian basement rocks was because of the prevalence of major shear zones. Examples of these structures were described in the years following Watterson's paper in a number of Precambrian regions including northwest Scotland and southern Africa as well as in southern Greenland.

Structures of the kind represented by Watterson's major F3 fold are examples of ductile shear zones, and in the Vesterland case, the wide zone of mylonites and mylonitic gneiss is the central most highly deformed part of such a zone. That being so, the Flinn homogeneous strain ellipsoid cannot apply to such a structure, since the Flinn model strictly applies to irrotational strain,

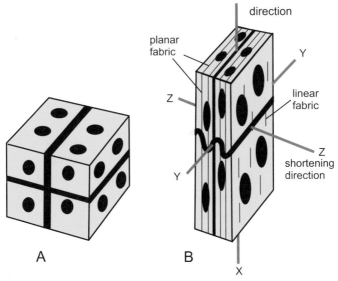

Figure 5.8 Strain and fabric. Block diagram to show how various structural elements (planar layers and spheres) in a deformed cube combine to give a picture of the overall strain. The initially vertical layer is extended parallel to X and thinned, while the horizontal layer is folded parallel to Z. The spheres are transformed into ellipsoids, which are elongated parallel to X and flattened parallel to Z. The deformed body acquires a fabric with a planar element parallel to XY and a linear element parallel to X.

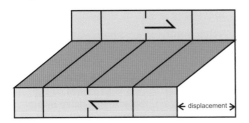

A homogeneous simple shear

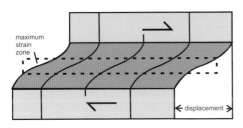

B heterogeneous simple shear

Figure 5.9 Shear zone geometry in two dimensions (i.e. plane strain): a zone of ductile strain between two undeformed blocks moving in opposite directions. **A** In homogeneous simple shear, the shear strain is uniform across the zone but there is an abrupt change at the margins. **B** In heterogeneous simple shear, the shear strain varies from a minimum at the margins to a maximum in the centre of the shear zone.

whereas in an ideal zone of simple shear, the strain is rotational and heterogeneous, varying continuously towards a maximum value in the central part of the zone (Fig. 5.9).

In terms of displacement, the Vesterland shear zone represents the deep-crustal equivalent of a strike-slip, or wrench, fault. The estimated values for the amount of strain calculated by Watterson would therefore apply to the highly strained parts of the F3 fold, close to the mylonite belt, but the strain in the most highly strained central part of the zone would not have been measurable. Watterson did however recognize one of the essential characteristics of shear zones, which is that the F3 deformation was accomplished by shear, or flow, sub-parallel to the axial plane of the F3 fold, rather than by buckling, for example, and that much of the pre-existing structure had been brought into near-parallelism with it. Thus the XY plane approximates to the shear plane and the X direction to the shear direction.

In a later paper, published in 1974 with A. Escher, Watterson's model was extended to include rotational simple shear strain. The authors point out that zones of simple shear are the only way of explaining the relatively rapid transition between unstrained or weakly strained rocks and the extremely high strains that are found commonly in narrow belts within basement terrains. As the authors state:

> 'The observed continuity of lithological horizons across deformation boundaries cannot be accounted for in any other way .... Boundaries between undeformed rocks and rocks subjected to pure shear [i.e. irrotational] deformation must inevitably result in discontinuities [e.g. faults].' [Figure 5.10 explains this point.]

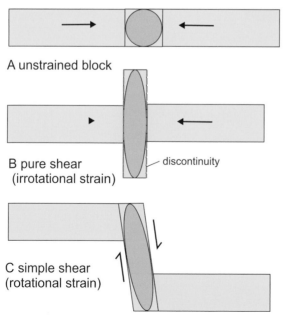

A unstrained block

B pure shear
(irrotational strain)

discontinuity

C simple shear
(rotational strain)

Figure 5.10 Sketch to show how large strains cannot be accommodated by irrotational strain (**B**) without discontinuities intervening, whereas zones of simple shear (rotational strain) (**C**) can.

This insight, which seems obvious now, was the key to understanding shear zones.

However, the first attempt to provide a rigorous mathematical analysis of a shear zone was published by John Ramsay and Rod Graham in 1970, Ramsay providing the mathematical background and Graham an account of three actual examples: two from the Swiss Alps and one from the Lewisian of Northwest Scotland. In their conclusions, they write:

> 'if the walls of the shear zone are undeformed and volume change is unimportant, such zones can only be formed by the process of heterogeneous simple shear. Shear zones formed in this way are particularly important, because their kinematic significance, states of strain, and displacement differences across the zone may be calculated.'

And further:

> 'Shear zones are common features in the deformed crust, and these new techniques might have considerable application in the investigation of large- and small-scale crustal displacements.'

All displacements across major crustal or lithospheric boundaries, such as subduction zones and major intra-crustal sutures, involve a translation by means of a shearing mechanism such as a thrust or strike-slip fault movement. At depth, such displacements must be represented by their ductile equivalents, which are shear zones (Fig. 5.11). Ramsay and Graham demonstrated how the analysis of the strain variation across such zones could yield an estimate of the amount of displacement, and was a most valuable addition to the tools available to geologists attempting to reconstruct former Earth movements. A subsequent paper by Ramsay in 1980 presents a comprehensive review of shear zone geometry, summarizing the research on the subject up to that date.

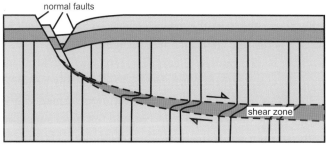

normal faults

shear zone

Figure 5.11 A displacement accomplished by brittle fault-ing in the upper crust becomes a ductile shear zone at depth. After Ramsay, 1980.

Shear zone geometry

The main elements of shear-zone geometry are summarized in Figure 5.12. Assuming that the displacement took place under conditions of plane strain (i.e. behaved in the same way as a fault), the geometry is essentially two-dimensional.

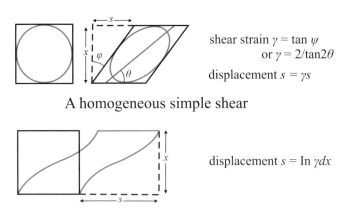

shear strain $\gamma = \tan \psi$

or $\gamma = 2/\tan 2\theta$

displacement $s = \gamma s$

A homogeneous simple shear

displacement $s = \text{In } \gamma dx$

B heterogeneous simple shear

Figure 5.12 Measurement of shear strain and displacement in plane strain. **A** In homogeneous simple shear, the shear strain is uniform across the zone and the shear strain and displacement are calculated as simple functions of the angle ψ made by a formerly perpendicular line and the angle θ made by the long axis of an ellipse formed by a circle in the undeformed block. **B** In heterogeneous simple shear, the displacement is the sum of the incremental displacements across the zone.

If the rock adjacent to a planar zone of ductile shear is undeformed, and volume change is discounted, the displacement across the zone can be calculated according to simple geometric rules assuming that the deformation has been accomplished by a process of progressive simple shear.

In the simple case of homogeneous simple shear (Fig. 5.12A), the shear strain γ is measured by the change in the angle ψ, the deflection of the sides of an initial square, or the angle θ made by the X axis of the deformation ellipse formed by the distortion of an original circle. In a natural ductile shear zone, the strain does not normally increase suddenly to a maximum from zero at the margin of the zone, and the strain is heterogeneous. However, if the strain varies smoothly across the zone the displacement can be calculated by summing the values of γ across the zone as shown in Figure 5.12B.

Various methods have been proposed to calculate strain and displacement in naturally occurring shear zones. In zones developed in originally homogeneous rocks such as igneous bodies (Fig. 5.13), a foliation is formed that is parallel to the XY plane of the deformation ellipsoid, and this gives a method of analysing the strain variation across the zone. The foliation (e.g. cleavage or schistosity) is formed initially at an angle of 45° to the shear zone boundary, but the angle becomes progressively smaller towards the most highly strained centre of the zone.

Planes and lines within the undeformed rock are also deflected across the shear zone boundary, and this gives another way of estimating displacement

shear zone

Figure 5.13 Shear zone in previously undeformed granite. Note that there is a faint foliation within the granite that curves into the central high-strain zone.

across the zone. Because the undeformed structures may have a variety of orientations prior to the deformation, the calculation of strain using this method is complex, and it has been pointed out that spurious or conflicting results can be obtained merely by observing the apparent displacement of planar objects, which depends on their initial orientation. A mathematical solution to the problem of the rotation and deformation of randomly oriented planes and linear structures during progressive simple shear was published by Lilian Skjernaa in 1980 and the problem was also discussed by Ramsay in the same year.

Movement indicators

The Ramsay–Graham analysis can only be applied to shear zones where the strain varies smoothly across the zone and where this variation can be easily measured, (e.g. by the shapes of deformed objects, etc.). Many shear zones are either more complex – they may include a brittle component – or the strain variation may be too variable. In such cases, it is useful to be able at least to determine the sense of movement across the zone (i.e. either sinistral or dextral in the case of strike-slip zones). Many ingenious methods have been proposed, some of which are illustrated in Figure 5.14.

Strongly developed linear structures are a prominent feature of many large-scale shear zones and are a reliable indicator of the direction of movement

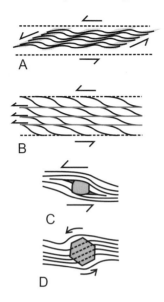

Figure 5.14 Sense of movement indicators in shear zones. **A** Extensional crenulation cleavage bands (orange layers): the smaller angle with the main foliation points in the direction of movement. **B** The sense of movement on secondary shears (in red) is parallel to the shear sense on the main zone. **C** *Sigma structure*: objects such as porphyroblasts develop recrystallization 'tails' resembling the Greek letter *sigma* (σ) that point in the direction of movement. **D** The sense of rotation in a porphyroblast is indicated by the rotation of an internal fabric.

across the zone. Shackleton and Ries pointed out in 1984 that regional stretching lineations give a reliable method of determining the relative motion of converging plates, using examples from the Himalayas and the Variscan and Caledonian belts of Europe. This technique has now become routine.

Postscript

As a direct result of the Ramsay–Graham study, the investigation of shear zones became a major focus of structural geologists worldwide, prompted by a group of Imperial College students working on the Lewisian complex of NW Scotland. However, it was the pioneering efforts of Flinn and Watterson that provided the initial stimulus.

The work of these geologists enabled the significance of shear zones as sites of major crustal displacement to be recognized, and attempts could then be made to discover the direction and sense of movement across them. Such zones in many cases represent important former plate boundaries, or intra-plate displacements, whose kinematic behaviour offered an important means of reconstructing plate movement patterns in the geological past.

6

Plate tectonics

Historical background

There can be little doubt that the introduction of the theory of plate tectonics in the late 1960s has been the single most influential breakthrough in the earth sciences since the time of Darwin. By around 1960, the studies of palaeomagnetic pole positions from various continents for different geological periods had finally convinced the doubters that continental drift was a valid theory (see chapter 3). At the same time, as a result of the work of H.H. Hess and others on the nature of the ocean floors, it had become generally accepted among geologists that the movements of the continents were controlled by lateral movements of the oceanic crust, and that these themselves were part of a mantle-wide system of solid-flow convection, as outlined in chapter 4 (see Fig. 4.4). What was then needed was a conceptual model that explained how this process worked in detail; this model came to be known as plate tectonics, and revolutionized the earth sciences because it provided an explanation for so much that had hitherto been unexplained or insufficiently understood.

The plate-tectonic theory evolved gradually out of the concept of continental drift thanks to the insights of several influential figures, prominent among whom were H.H. Hess and J.T. Wilson. Hess's contribution arose from oceanographic work on the ocean floors and gave rise to his hypothesis of 'sea-floor spreading'.

Sea-floor spreading
Topography of the ocean floor
Extensive oceanographic mapping by various remote-sensing techniques had revealed a topography that was as varied as that of the continents. It became clear that the generally even ocean floor is interrupted by a system of great ridges and deep, narrow trenches (Fig. 6.1). The ridges are typically between 1000 and 2000 kilometres wide and rise to as much as 3 kilometres from the ocean floor. They make a continuous network, one branch of which runs from the Arctic along the centre of the Atlantic Ocean (the mid-Atlantic ridge) to

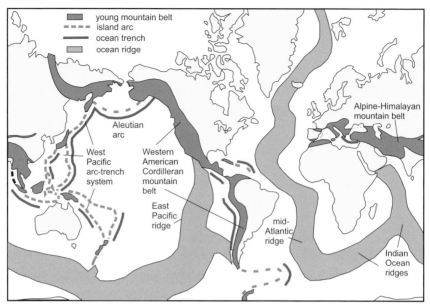

Figure 6.1 Principal topographic features of the Earth. Note the size of the ocean ridges compared with the continental mountain belts and the narrow ocean trenches. After Wyllie, 1976.

join a second branch that partly encircles Antarctica and crosses the Pacific Ocean towards the coast of Mexico; a third branch extends north into the Arabian Sea. The trenches are much narrower (typically around 100 kilometres wide) but extend to depths of up to 11 kilometres below sea level. They form generally curved linear features on the ocean-ward side of island chains around the western Pacific Ocean, the eastern Indian Ocean, and the Caribbean, and fringe the Pacific coasts of Central and South America.

Much of this information had been available even in Wegener's time, but its significance was not understood until the sea-floor spreading idea at last gave a believable explanation. Extensive work by American oceanographers based at Scripps Institute in California and Lamont Geological Observatory at Columbia University, New York, produced a greatly expanded knowledge of the topographic detail of the ridge–trench system.

The contribution of H.H. Hess

Harry Hammond Hess (1906–1969) was one of the most influential geologists of the twentieth century. Hess was a graduate of Princeton University, and after accompanying Vening Meinesz in his submarine-based gravity survey of the East Indies in 1932, returned to Princeton as a member of staff in 1934. During his service in the US Navy in World War II, Hess used the then novel sonar technology to survey the floor of the north Pacific, and discovered the flat-topped Pacific seamounts, which he termed 'guyots'. His theory of how these volcanic islands developed is discussed in chapter 10.

Sea-floor spreading

The essence of Hess's proposal is that the ridges are underlain by hot rising mantle currents and the deep ocean trenches by cool descending currents; he called this process 'sea-floor spreading'. It proved to be the major breakthrough in understanding that finally led to a solution to the problem of a mechanism for continental drift. First published in a Report of the (US) Office of Naval Research in 1960, his theory was explained in detail in an article entitled *History of ocean basins* in1962; similar ideas were expressed by R.S. Dietz in Runcorn's 1962 book on *Continental Drift* referenced in chapter 3.

The significance of the guyots

Hess notes that a broad band of guyots crosses the Pacific Ocean from Chile to the Marianas and stops abruptly at the Mariana Trench. Their flat tops indicate that they have been planed down by erosion, and in some cases they are overlain by Cretaceous or Eocene shallow-water sediments. The fact that they now lie well below sea level implies that the sea floor has subsided since their formation (Fig. 6.2). Moreover, their comparative geological youth indicated that the Pacific Ocean was not ancient as Wegener had thought, and proved that the ocean crust must be continuously renewed. He concluded:

> 'the whole ocean is virtually swept clean (replaced by new mantle material) every 300 to 400 million years.'

In fact, as we now know, none of the oceanic crust is older than Triassic.

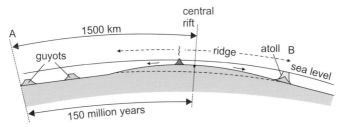

Figure 6.2 Hess's model of guyot formation. Volcanoes form near the ridge axis, become planed down to sea level by erosion, then move sideways away from the axis as new crust is created and subsides as the crust cools. At point A, the guyot has moved 1500 km in 150 million years. The old volcano at point B has hosted a coral reef that has continued to maintain its surface near sea level as it subsides. After Hess, 1962.

The significance of the ridges

Hess observes that ocean ridges exhibit high heat flow and shallow seismicity along a central rift-like structure, suggesting extension perpendicular to the ridge axis; that sediment cover is thin or absent, and that the Mohorovičić discontinuity (the seismically determined base of the crust – see chapter 4) cannot be detected there. He suggests that the extra height of the ocean ridges compared to the rest of the ocean floor is due to their lower density, caused by the extra heat provided by a rising mantle current. The subsidence of the guyots is attributed to sinking of the sea floor due to cooling as it moves away from the ridge.

Some of Hess's main conclusions are worth repeating verbatim:

- 'The mid-ocean ridges could represent the traces of the rising limbs of convection cells while the circum-Pacific belt of deformation and volcanism represents descending limbs.'
- 'The continents do not plough through oceanic crust impelled by unknown forces; rather they ride passively on mantle material as it comes to the surface at the crest of the ridge and then moves laterally away from it.'
- 'Their leading edges [of the continents (author)] are strongly deformed when they impinge upon the downward moving limbs of convecting mantle.'
- 'The cover of oceanic sediments and the volcanic seamounts also ride down into the jaw crusher of the descending limb, are metamorphosed, and probably welded onto continents.'

Much of the objection to Wegener's ideas on continental drift had centred on the failure to visualize how a continent could plough across a static ocean crust. However, this objection was countered by Hess's proposal that the ocean crust itself was mobile, rising at the ridges and moving sideways towards the deep ocean trenches where it descended. In other words, both continents and oceans were mobile rather than static. This idea differed from the Holmes model (see Fig. 4.4) in visualizing the oceanic crust as part of the circulating layer, rather than sitting above it, as Holmes had conceived it. Moreover, the new model incorporated the recently introduced concept of the 'lithosphere', which included the crust and was much thicker than it. Thus the oceanic lithosphere was regarded by Hess as a giant moving 'conveyor belt', being continuously renewed at ridges and thereby transporting the continents around the Earth's surface (Fig. 6.3). The elegant model of a mid-ocean ridge provided by Vine and Matthews in 1963, using the new palaeomagnetic data (see chapter 7 and Fig. 7.3) provided the necessary proof that the sea-floor spreading hypothesis was valid.

Hess called his essay on the history of ocean basins 'geopoetry', and although some of his ideas (e.g. that the oceanic crust consisted of serpentinized peridotite) have been subsequently rejected, there is no doubt that his ideas were a major influence on the development of plate tectonics.

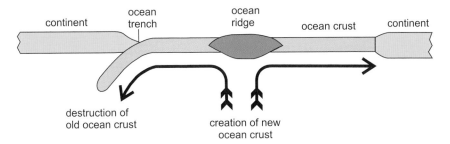

Figure 6.3 Hess's model of sea-floor spreading. New ocean crust is created at ocean ridges and destroyed at ocean trenches by flow within the mantle.

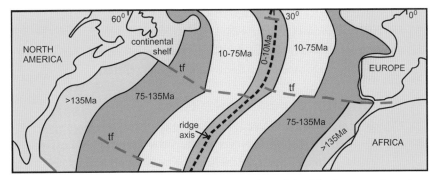

Figure 6.4 Dating the ocean floor. The age distribution of the Atlantic ocean floor given by palaeomagnetic measurements displays a simple pattern where the youngest dates (0–10 Ma) are in a central strip (red) along the axis of the mid-Atlantic ridge. On each side of this central strip are successively older bands with the oldest (>135 Ma) adjacent to the North American and African coasts respectively. These dates correspond to the dates when these continents originally separated. Note that north of the transform fault, the oldest dates are younger (75–135 Ma) than those to the south, indicating that Europe separated from North America at a later date. tf, transform fault. Based on Larson and Pitman, 1972.

Dating of the ocean floor

Wegener, Du Toit and Holmes had all thought that the ocean ridges marked the lines of separation of the continents, and consisted of foundered continental crust (see Fig. 4.4), but the palaeomagnetic dating of the ocean floor of the Atlantic and Indian oceans in the 1960s showed that the ridges were the most recently formed parts, and that the ocean floor became older towards the continental margins.

The age pattern of the Atlantic ocean floor given by palaeomagnetic measurements (Fig. 6.4) indicates a simple pattern where the youngest dates (0–10 Ma) are in a central strip (red) along the axis of the mid-Atlantic ridge, proving that the ridge is the site of formation of new ocean crust at the present day. On each side of this central strip are successively older bands with the oldest (>135 Ma) adjacent to the North American and African coasts respectively. These dates thus correspond to the dates when these continents originally separated. Note that north of the fault separating Africa from Europe, the oldest dates are younger (75–135 Ma) than those to the south, indicating that Europe separated from North America at a later date. This work finally proved that the continents of Gondwana and Laurasia had indeed moved apart and that the space between had been filled by new ocean crust.

Ocean crust could not be continuously created without it being destroyed elsewhere (assuming the Earth was not expanding!), and the obvious sites for its destruction were the deep ocean trenches as proposed by Hess. The new palaeomagnetic dating evidence demonstrated that the ocean crust adjacent to the trenches showed a variety of ages (i.e. a 'discordant' age pattern) as will be seen later (see Fig. 6.9), whereas the crust adjacent to continental margins that had moved apart showed a 'concordant' age pattern, consistent with the date of separation, as shown in Figure 6.4. This evidence confirmed that the Hess conveyor belt model for the ocean floor was essentially correct.

The significance of earthquake zones

The plate-tectonic theory evolved gradually out of the sea-floor spreading concept thanks to the insights of several influential figures. Foremost among these was Tuzo Wilson, of Toronto University, who recognized the significance of the linear zones of volcanoes and earthquakes that delineate a worldwide, tectonically active, network. Figure 6.5 shows that the sites of the major earth-quakes follow a network of narrow zones along the crests of the ocean ridges

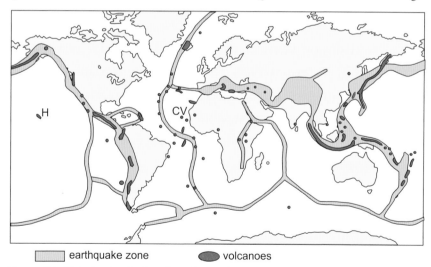

earthquake zone   volcanoes

Figure 6.5 Global network of earthquakes and volcanoes. The earthquake zones follow the tectonically active crests of the ocean ridges, the island arcs of the western Pacific and the Alpine–Himalayan and American Cordilleran mountain chains. Most volcanoes are concentrated along the same zones but some (e.g. Hawaii, H and Cape Verde, CV) also occur within the ocean basins and continents. After Chadwick, 1972.

and within broader and less well-defined zones along the island-arc–trench system of the Western Pacific and Indian Oceans, and the orogenic belts of the continents. The great majority of active volcanoes also occur within the same belts, although a few occur within the ocean basins, such as the Hawaiian islands in the Pacific and the Cape Verde archipelago in the Atlantic.

The Benioff-Wadati zones

The distribution of earthquake foci was also used to demonstrate what happens to the oceanic crust at the deep trenches that exist around much of the Pacific Ocean. Two seismologists, Hugo Benioff of the California Institute of Technology and Kiyoo Wadati of the Japan Meteorological Agency, had independently discovered that deep-focus earthquakes along parts of the volcanic island arc–trench network fell on a plane that dipped at an angle of around 45° beneath the line of the trench, and these zones became known as Benioff or Benioff-Wadati zones (Fig. 6.6) and subsequently as 'subduction zones' (Benioff, 1949). Detailed studies of the earthquakes in these zones were thus able to define not only the direction of movement but also the angle of inclination at the earthquake focus and confirmed Hess's earlier suggestion of a downward movement of the oceanic crust along the trench network.

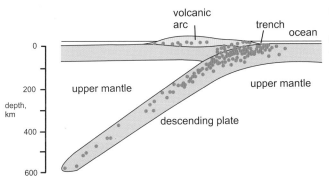

Figure 6.6 The Benioff–Wadati zone. The deep earthquakes (red dots) lie in a plane inclined beneath the volcanic island arc. Shallow earthquakes occur both on the volcanic arc and on the descending plate. After Benioff, 1949.

The contribution of Tuzo Wilson

J. Tuzo Wilson (1908–1993) was born in Ottawa and graduated in geophysics in 1930 at the University of Toronto. He then gained a BSc degree at Cambridge, and proceeded to Princeton to obtain his doctorate under Harry Hess. After serving in the Canadian Army during World War II, he returned to Toronto in 1946 as Professor of Geophysics. He also served as Director of the Royal Ontario Museum from 1974 to 1985. He received many honours including Fellowship of the Royal Societies of London, Edinburgh and Canada. Wilson's claim to be the progenitor of plate tectonics lies in two papers published respectively in

1963 and 1965: the earlier, entitled *Hypothesis of Earth's behaviour* and the later, *A new class of faults and their bearing on continental drift.*

Transform faults

Wilson realized that the concentration of tectonic activity indicated by the earthquake zones must define the places where crustal movements were taking place at present, and that the network of zones must therefore mark the boundaries of relatively stable aseismic blocks of crust whose margins could thus be traced by following the earthquake zones.

In his ground-breaking paper of 1965, Wilson pointed out that the numerous faults that appeared to offset the ridge axes and the trenches were also delineated by earthquakes, and were themselves part of the boundary of the stable blocks. He explains that the ridges, trenches and faults:

> '…are not isolated, that few [actually none! (author)] come to dead ends, but that they are connected into a continuous network of mobile belts about the Earth which divided the surface into several large rigid plates. Any feature at its apparent termination may be transformed into another feature of one of the other two types.'

This quotation is significant in containing what is probably the first mention of the term 'plate' in this context.

Wilson realized that these faults effectively changed the sense of movement across the boundary from, say, divergent across an ocean ridge to one that was parallel to the fault. He termed such faults 'transform faults' since they transformed the sense of motion (e.g. from divergent to strike-slip) along the boundary of a block. He stressed that transform faults so defined were not true transcurrent (or wrench) faults and should be distinguished from them.

Figure 6.7 illustrates Wilson's concept with two examples: in A, plates X and Y are moving apart and divergent motion across the ridge is transformed to motion parallel to the fault that joins the ends of the ridge segments; in B, plates X and Y are moving together (Y is being thrust beneath X) and convergent motion across the trench is converted into motion parallel to the transform fault. Note that in both A and B, the sense of relative motion along the fault is dextral, or right-lateral (i.e. for an observer standing on one side, the opposite side moves to the right). Figure 6.7C illustrates an example of a strike-slip, or wrench, fault, which is a common type of fault found on land. This has the same apparent sense of displacement as the transform fault in A and B, in that the opposite side moves to the left. However here, the whole block on one side moves in the same direction and the sense of movement is sinistral, or left-lateral. Thus an important property of a transform fault is that the apparent movement sense (in this case sinistral) is the opposite of the

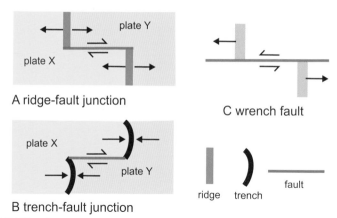

A ridge-fault junction

B trench-fault junction

C wrench fault

ridge   trench

fault

Figure 6.7 Wilson's explanation of transform faults. **A** Plates X and Y are moving apart from a ridge, and the sense of movement along the transform fault is dextral – i.e. the opposite side is moving to the right. **B** Plates X and Y are moving together at a trench; the sense of movement along the fault is also dextral. **C** In a wrench fault, a green band on the opposite side of the fault is moved to the left and the sense of movement on the fault is the same as the movement of the green band – i.e. sinistral. Based on Wilson, 1965.

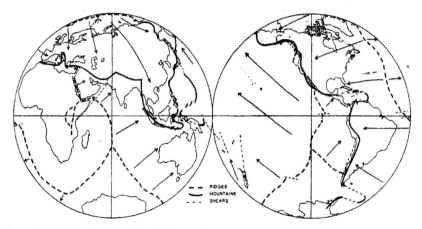

Figure 6.8 Wilson's map of World plates. Compare Figure 6.13. From Wilson, 1965, with permission.

actual sense, which in this case is dextral. So Wilson's discovery meant that many faults, including well-known examples of important faults that had previously been classed as wrench faults for many years, such as the San Andreas Fault of California, were re-classified by him as transform faults.

Wilson illustrates his theory with examples drawn from the central Atlantic (see Fig. 6.4), the northeast Pacific, and the De Geer fault, which terminates the North Atlantic ridge south of Svalbard; he illustrates his interpretation of the movement directions of the main plates in two sketch maps (Fig. 6.8).

The North-East Pacific

The significance of Wilson's introduction of the transform fault is well illustrated by the north-eastern Pacific region (Fig. 6.9) Here, the western boundary of the stable American Plate can be traced as an active earthquake zone from the Central America Trench in the south to its intersection with the East Pacific Ridge, from where it continues north-westwards as the San Andreas

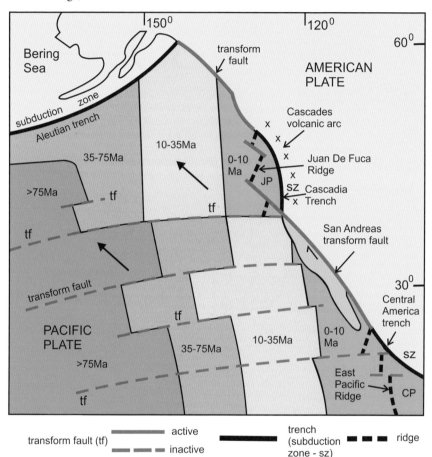

Figure 6.9 The NE Pacific. The boundary between the North American plate and the Pacific plate is defined by both subduction zones and transform faults (including the famous San Andreas Fault). In the Pacific plate the magnetic stripe pattern defines several dated zones which are offset by a series of transform faults. The East Pacific and Juan De Fuca ridges end against the San Andreas Fault and the movement direction changes from convergent across the subduction zones to margin-parallel along the fault. The transform faults within the Pacific are all parallel to each other and represent the direction of movement of the Pacific plate relative to the East Pacific ridge. In the north, the Pacific plate is being subducted beneath the Aleutian trench. Note that the movement direction of the Pacific plate relative to North America is NW-wards, parallel to the San Andreas Fault, and that the piece of North American continent west of that fault is also travelling north-westwards, relative to the rest of the continent. sz, subduction zone; tf, transform fault. Based on Larson and Pitman, 1972.

Transform Fault until it meets the Cascadia Trench offshore from the Cascades volcanic arc. At the northwest end of this trench, the boundary continues along the coast as another transform fault, ending at the Aleutian Trench off northwest Alaska. Each of these transform faults ends abruptly at a trench, at which point the sense of movement changes from one that is parallel to the fault to one that is convergent across the trench.

The Cascadia Trench sector exists because of a small oceanic plate (the Juan de Fuca Plate) situated offshore from the trench and separated from the main Pacific plate by a short section of ocean ridge, the Juan de Fuca Ridge. Within the Pacific Plate itself, all the transform faults are now inactive, the only currently active sections being those offsetting the active sections of the East Pacific and Juan de Fuca ridges. The inactive transform faults, which trend approximately E–W, are parallel to the former direction of relative motion between the Pacific Plate and the old Farallon Plate, now subducted beneath North America, whose modern representatives include the small Juan de Fuca and Cocos Plates.

Tectonics on a sphere

The final stage in the construction of the plate-tectonic model following from Wilson's theory was based on the proposition that both continental and oceanic crust must behave in a semi-rigid manner, moving laterally as single units or blocks, and that relative movement between the blocks was concentrated at their boundaries. Three scientists in particular were responsible for developing and testing the new plate model: Dan McKenzie of Cambridge, Jason Morgan of Princeton, and Xavier Le Pichon, at Lamont Geological Observatory (now Lamont-Doherty of Columbia University), who all appear to have developed their ideas more or less at the same time.

The contribution of McKenzie and Parker

The first of these contributions to appear in print was by Dan McKenzie of Cambridge in 1967, with R.L. Parker of the University of California at San Diego, who investigated the properties of plates, viewed as pieces of a spherical surface. They observed that evidence for the essential rigidity of the plates came from the fact that linear features on the ocean floor, such as the faults and the striped magnetic pattern, were essentially unaffected by warping or bending such as might be expected if the ocean floor were to behave in a 'plastic' manner. They note that linear magnetic anomalies can only be produced close to the ridge axis, and that:

> 'The spreading seafloor then carries these anomalies for great horizontal distances with little if any deformation.'

A further illustration of this is provided by the opposing coastlines of Africa and South-Central America, which still show a good fit despite having travelled away from each other for a distance of about 3000 kilometres over a period of 150 million years. Thus South America and the western half of the South Atlantic, on the one hand, and Africa and the eastern half of the South Atlantic on the other, can be considered as separate blocks, which moved as units. The Atlantic oceanic age pattern (see Fig. 6.4) confirms this.

After reviewing the evidence from the distribution of seismicity, the authors conclude:

- 'These observations are explained if the seafloor spreads as a rigid plate, and interacts with other plates in seismically active regions which also show recent tectonic activity...' [and]
- 'Ridges and trenches are respectively defined as lines along which crust is produced and destroyed...' [and further]
- 'Transform faults conserve crust and are lines of pure slip. They are always parallel to the relative velocity vector between two plates – a most useful property.'

The basic idea of the rigid plate is illustrated in Figure 6.10: three plates are shown, A, B and C. Plates A and B are moving apart away from a ridge, which is offset by two transform faults. Because the motion of the plate must be parallel to the faults, this determines the relative movement direction of A and B. On the left side of plate B, B is descending beneath C at a trench; this process is known as subduction and the trench marks the position of the subduction zone. Again, the movement direction of plate B relative to plate C must be parallel to the transform fault that offsets the subduction zone. This

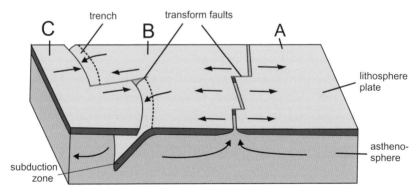

Figure 6.10 The lithosphere plate. Model showing three lithosphere plates, A, B and C. A and B are moving apart at a ridge offset by two transform faults. The movement direction is parallel to the transform faults. B is being subducted beneath C; here the movement direction is in places slightly oblique to the trench but is parallel to the transform faults. The arrows in the asthenosphere represent possible compensating flow there. After Isacks, Oliver and Sykes, 1968.

direction is oblique to the previous one – in other words, the direction of relative movement of plate B is different for each boundary.

Thus an important property of plates is that they may move obliquely to a ridge or a trench but must always be parallel to any transform faults at their boundaries. At first sight, it may look as if plate B is moving in two different directions simultaneously, but this is an illusion: both these movement arrows represent relative movements across different boundaries; this will be understood more easily if plate B is regarded as stationary; then plate A is moving away from it in one direction and plate C towards it in another direction. The principle that rigid plates can move obliquely to a convergent boundary was an important insight: it explained how mountain chains such as the Alps or Himalayas have curved, or even quite contorted, shapes yet are caused by the convergence of only two rigid plates whose movement directions must therefore be quite oblique to some sectors of the chains.

The rigid plates were conceived as being composed of pieces of the lithosphere, with a definite thickness determined by the depth at which the material of the Earth's mantle ceases to behave in a semi-rigid manner and becomes capable of slow plastic flow in the solid state. As will be described in chapter 7, the lithosphere consists of the crust, plus part of the upper mantle, and is underlain by the weaker 'asthenosphere'. The depth of the lithosphere varies, being thinner near the warm ridge and thicker at the cooler trench, but the mean thickness is around 100 km. It is the difference in properties between the lithosphere and the asthenosphere that enables the plates to move laterally over the weaker asthenosphere.

The authors point out that the plates must act as pieces of a spherical shell and that all plate movements can be understood in terms of rotations:

> 'If one of two plates is taken to be fixed, the movement of the other corresponds to a rotation about some pole [i.e. axis (author)] and that all relative velocity vectors between the two plates must lie along small circles or latitudes with respect to that pole. If these small circles cross the line of contact between two plates, they must be either a ridge or a trench but if it is itself a small circle, it must be a transform fault.'

McKenzie and Parker apply this reasoning to the north-eastern Pacific Plate, noting that the transform faults within the plate are all parallel and slightly curved (see Fig. 6.9). They showed that the motion of the Pacific Plate relative to the East Pacific Ridge was westwards, parallel to these transform faults, whereas its motion relative to the North American Plate was north-westwards, parallel to the San Andreas Fault, and towards the Aleutian subduction zone (i.e. in the direction of the arrow on Figure 6.9).

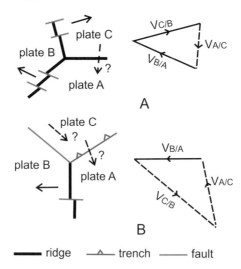

Figure 6.11 Determination of plate movement vectors in the case of three plates meeting at a 'triple point'. McKenzie and Parker's method of finding the direction and magnitude of movement of a third plate, given the directions and magnitudes of the other two by means of a vector triangle. In **A**, three ridges meet at a triple point: the velocity of B with respect to A, VB/A, is given by the spreading rate and the direction of the transform fault, and the velocity of C with respect to B, VC/B, is determined in the same way, then the velocity of A relative to C, (VA/C) is given by the third side of the vector triangle. In case **B**, of a ridge, trench and fault meeting at a triple point, the velocity of B relative to A is obtained in the same way as for case A. The direction of C relative to B is given by the transform fault, so that the vector triangle can be completed if either the spreading rate of C relative to B or the direction of A relative to C can be found. After McKenzie & Parker, 1967.

The authors also point out that the plate velocities can be obtained from ocean ridge spreading rates, corrected by calculating their value parallel to the transform faults that cut them. Figure 6.11 demonstrates their method of obtaining an unknown plate movement vector in the case of three plates meeting at a 'triple junction'. In case A, of three ridges meeting at a triple point, the velocities of B relative to A and C relative to B are given by the spreading rates on the two ridges and the directions of the transform faults. The velocity of A relative to C is then obtained by completing the vector triangle. In case B, of a ridge, trench and fault meeting at a triple point, the velocity of B relative to A is obtained in the same way as for case A. The direction of C relative to B is given by the transform fault, so the triangle can be completed if either the spreading rate of C relative to B is known, or the direction of A relative to C can be found from a transform fault cutting the trench. By applying this method globally, all the plate vectors could be found.

Movement vectors obtained for the North Pacific were consistent with those obtained from earthquakes around the West Pacific island arcs, and the authors conclude:

'The North Pacific shows the remarkable success of the paving stone [sic] theory over a quarter of the Earth's surface, and it is therefore expected to apply to the other three-quarters.'

The contribution of Jason Morgan

Morgan had the same idea of transferring the plate movements onto a spherical surface, having also noted that the transform faults, especially in the oceans, were usually curved. He applied the method on a global scale, dividing the Earth's crust into twenty 'blocks', all assumed to be perfectly rigid, whose relative motions could be described by rotations about a single axis emerging somewhere on the Earth's surface as a pole (Fig. 6.12). Morgan noted that the transform faults in the Central Atlantic Ocean approximated to small circles whose centre was located near the southern tip of Greenland. He also discovered that the Atlantic Ocean spreading rates varied from a maximum in the central Atlantic to a minimum north of Iceland and concluded that both these sets of data were consistent with motion about a single axis of rotation. His paper was published in 1968.

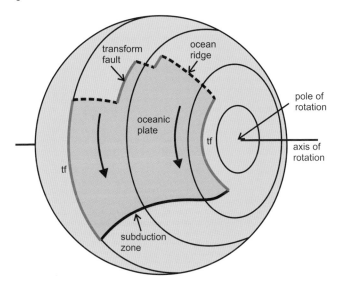

Figure 6.12 Plate tectonics on a sphere. The model illustrates the principle of plate tectonics on a spherical surface. The oceanic plate (green) is moving away from an ocean ridge and is being subducted beneath a trench. The movement direction is parallel to the transform faults that offset the ridge and also from the boundaries of the oceanic plate. The movement takes place around an axis of rotation that intersects the surface at the pole of rotation; the transform faults describe arcs of small circles about this pole. tf, transform fault. After Dewey, 1972.

Le Pichon and the six major plates

Xavier Le Pichon's contribution, also published in 1968, was to simplify the global plate structure into just six major plates: American, Eurasian, African, Indian, Antarctic and Pacific, together with a small number of minor ones (Fig. 6.13). The Indian plate has subsequently been divided into Indian and Australian, as shown in the figure. He showed that the movements responsible for opening all the major oceans – the Atlantic, Indian, Arctic, and the North and South Pacific – could each be explained by rotations about a single axis. The poles for these axes were found by the two independent methods: first by measuring the variation in spreading rates across the ocean ridges, as determined by the ages of the magnetic stripes, and secondly by plotting the intersections of the radii of the transform faults (i.e. the centres of the small circles made by the faults).

The spreading rates he measured varied from around 2 cm per year in the North Atlantic to more than 6 cm per year in the central Pacific. By combining the data, it was possible to compute the convergence rates across the trenches: 6 cm per year at the Peru Trench, 5–11 cm per year across the trenches of the Western Pacific, and 5.6 cm per year across the Himalayas.

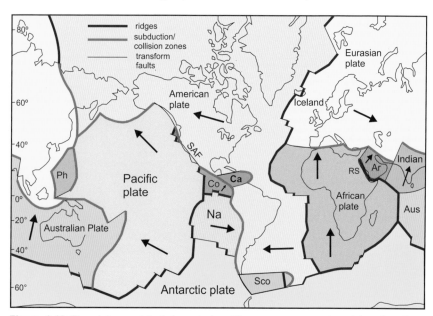

Figure 6.13 The plates and their boundaries. There are seven major plates: Americas, Eurasian, African, Indian, Australian, Pacific and Antarctic, together with several minor ones: Arabian (Ar), Cocos (Co), Caribbean (Ca), Nazca (Na), Philippines (Ph) and Scotia (Sco). The plates are separated by three types of boundary: constructive (ridges), destructive (trenches and collision belts) and conservative (transform faults). SAF, San Andreas Fault. The arrows give the direction of motion of each plate relative to the Antarctic Plate (regarded as stationary). After Vine and Hess, 1970.

A seismological test of plate tectonics

An influential paper by Isacks, Oliver and Sykes, also published in 1968, summarizes the wealth of seismological data held or published up to that point in order to test whether, and to what extent, it supported the new plate-tectonic theory. They refer to the world map of seismicity resulting from the compilation of around 29,000 earthquakes reported by the US Coast and Geodetic Survey between 1961 and 1967 by Barazangi and Dorman (1968) and conclude:

> 'Study of world seismicity shows that most earthquakes are confined to narrow continuous belts that bound large stable areas' [and that] 'Seismic phenomena are generally explained as the result of interactions and other processes at or near the edges of a few large mobile plates of lithosphere that spread apart at the ocean ridges... slide past one another along the large strike-slip faults, and converge at the island arcs and arc-like structures where surficial (i.e. crustal) materials descend.'

They note that in zones of divergence and strike-slip motion, seismic activity is moderate and shallow, whereas at island arcs and along active continental orogenic belts, seismic activity is much greater, more intense, and spread over a wider zone.

Fault-plane solutions

Analysis of the directions of first motion at the earthquake focus (termed the 'fault-plane solution') can reveal the orientation of the stress field responsible for the earthquake: whether the greatest stress direction was extensional or compressional, or whether the movement was essentially strike-slip.

The authors observe, firstly, that the solutions for the mid-ocean ridges are consistent with extension perpendicular to the trend of the ridges, as would be expected. Secondly, the fault-plane solutions for the transform faults confirm their strike-slip sense of movement and, critically, indicate the sense of movement as would be predicted by Wilson's transform theory rather than the opposite sense, which would have been appropriate for a traditional strike-slip fault on land (see Fig. 6.7).

In the case of the island arcs, the situation was more complex. Here, the seismicity is concentrated into three separate groups: a relatively shallow group of very active seismicity indicating compression parallel to the dip of the down-going slab; a second, of rather less active extensional activity, at the outer margin of the slab, which they attribute to the effect of the bending; and a third cluster of deep and intermediate-depth earthquakes indicating compression parallel to the dip of the slab. The authors conclude that the data are consistent with the concept of a relatively strong, cool, lithosphere slab moving through a warmer, more ductile asthenosphere. The shallow extensional events on the outer wall of the slab are consistent with the formation of extensional graben on the surface of the slab as it bends downwards, and the shallow compressional events with the resistance experienced by the down-going slab as it encounters stronger material at depth. The origin of the third deep-focus group has been the subject of debate; the authors suggest that the most likely explanation is that the seismicity results from mineralogical phase changes in the lithospheric material as it descends through deeper levels of the mantle. This topic is explored in greater detail in chapter 9 (see Fig. 9.7).

The authors also note that the slip vectors obtained from the focal plane solutions on individual oceanic plates are parallel to each other and also to the transform faults, supporting the central idea of plate tectonics that all parts of a plate should move in the same direction.

The example of the Gulf of Aden

Isacks *et al.* use the seismicity pattern in the Gulf of Aden, which is a young ridge offset by transform faults (Fig. 6.14), as an example of the use of seismicity to authenticate the plate model. The authors point out that the pattern of offsets on the ridge matches the stepped shape of the margins of the continents on either side of the Gulf, indicating that the 'faults' are not true faults whose offsets are due to fault movement, but must have existed as parts of a plate boundary since the inception of the rift.

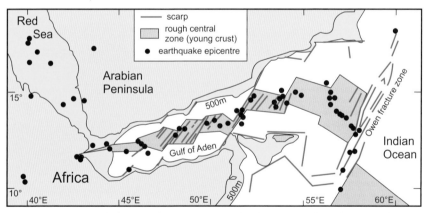

Figure 6.14 The Gulf of Aden rift zone indicated by the earthquake epicentres, showing the young central zone of new crust offset by transform faults, marked by escarpments. Continental shelf shown in yellow. Note that the transform offsets are matched by steps in the continental margins. After Isacks *et al.*, 1968.

Later refinements of the theory

As a result of the introduction of the plate-tectonic theory in the period 1967–68, there was a worldwide explosion of interest in it, and numerous geologists began applying it to explain many kinds of previously puzzling geological features. However, there shortly emerged a consensus over the basic elements of the plate theory.

Plate boundaries

The stable tectonic plates are completely surrounded by boundaries, which, as we have seen, are of three types: ridges, trenches and faults. Since the ridges must mark the sites where new ocean crust is produced, they have become known as 'constructive boundaries', and since the trenches mark the sites of destruction of ocean crust, they are termed 'destructive boundaries'. The transform faults correspond to zones where one plate merely slides past its neighbour without either creation or destruction of crust taking place. Consequently, because plate is 'conserved' at such boundaries, they are known as 'conservative boundaries'.

Constructive boundaries

These are of two types: oceanic spreading ridges and continental rift zones. The structure of a typical spreading ridge will be described in more detail in the following chapter, which explains how the creation of new oceanic lithosphere takes place (see Fig. 7.8). Some of the new material is in the form of basaltic magma produced by melting of the upper mantle, and is either injected into the crust in the form of intrusive bodies or poured out at the surface as lava flows. These rocks form the new oceanic crust; beneath it, new mantle lithosphere is formed by the addition of cooled material transferred from the asthenosphere. As new material is added, earlier-formed lithosphere moves sideways, cools, and gradually sinks to the normal level of the deep ocean floor.

Continental rifts

The Red Sea–Gulf of Aden–African rift system (Fig. 6.15) is an example of how a new constructive plate boundary may form in continental lithosphere. The three rifts join in a triple junction; in two of the arms of the system – the Red Sea and the Gulf of Aden – new oceanic crust has formed along the centre of the rifts and joins up with the Carlsberg ridge, part of the Indian Ocean ridge system. These new constructive boundaries define a separate Arabian plate, which is moving north-eastwards away from Africa. Both these rifts are destined eventually to become oceans as Arabia and Africa move apart. In this way the continents of the Americas, Eurasia and Africa would have separated during the split-up of Pangaea.

The African rift system can be taken as an example of how a spreading ridge might start. Its geological history is well documented: an uplifted region was

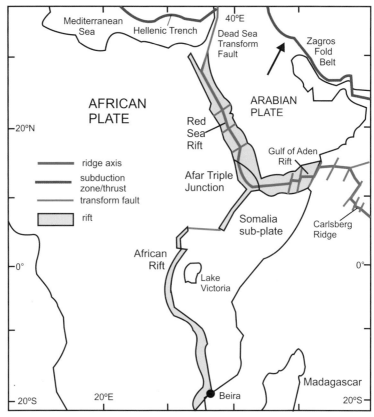

Figure 6.15 The Red Sea–Gulf of Aden–African Rift system. The Red Sea Rift joins the Gulf of Aden Rift at the northern end of the great African Rift system in northern Ethiopia. The Gulf of Aden Rift joins the NW end of the Carlsberg Ridge to form a continuous western boundary defining the separate Arabian plate, which is moving NE-wards away from the African plate. Oceanic crust has formed in the Gulf of Aden and Red Sea rifts, which are moving apart. Extensional movement has taken place across the African rift, which is recognized as the boundary of the small Somalia sub-plate. Note that only the currently active parts of the rift system are shown. After Girdler and Darracott, 1972.

formed, similar to an ocean ridge, with a central rift valley into which were poured the lavas that resulted from the melting taking place beneath. This was followed by a collapse of the rifts, accompanied by extensive volcanicity, but unlike the other two rifts, there has been no significant separation.

Destructive boundaries: subduction zones

There are three types of destructive plate boundary (Fig. 6.16). The first two types follow a zone of destruction of oceanic crust and are marked at the present day by the deep-ocean trenches; the third type is a zone of collision of two continental plates, and is represented by a belt of young mountain ranges such as the Alps and the Himalayas. This third type is, in geological terms,

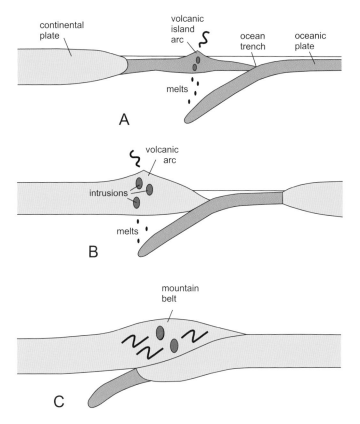

Figure 6.16 Destructive plate boundaries. **A** Oceanic subduction zone: oceanic lithosphere descends beneath another oceanic plate at a trench; melting occurs above the down-going slab, forming a volcanic arc on the upper plate. **B** Continent-margin subduction zone: oceanic lithosphere descends beneath the margin of a continental plate, forming a volcanic arc within the continental crust. **C** Continental collision zone: as two continents on opposing plates meet, the continent on the subducting plate is underthrust beneath the continent on the upper plate; the continental crust is thickened and deformed and further convergence eventually stops.

more short-lived, since the convergent movement of continental crust will eventually cease as the two continental plates grind together and gradually come to a halt.

The first two types of destructive boundary, therefore, are where oceanic crust belonging to one plate descends (i.e. is subducted) beneath another plate, which may be either oceanic (Fig. 6.16A) or continental (Fig. 6.16B). The lower of the two plates is always oceanic, and the deep-ocean trenches mark the subduction zones where the oceanic plates commence their descent; such zones are typically inclined beneath the upper plate at an angle that can vary from quite shallow to nearly vertical, depending on several factors, such as the rate of convergence and the age of the descending plate.

As the lower plate descends into warmer regions of the mantle, volatiles, especially water, are driven out of the crustal material and ascend into the mantle of the upper plate, part of which melts; the resulting magmas form igneous intrusions within the crust of the upper plate and feed volcanoes at the surface. Where the upper plate is oceanic, as in Figure 6.16A, the volcanoes are partly submerged and form an island arc.

In a typical island arc, the partially submerged volcanic mountain range, which may be 50 to 100 kilometres wide, is bounded on its outer side by a trench situated between 50 and 150 kilometres from the volcanic arc. Between the arc and the trench is a zone where volcanic material accumulates, together with sediments resulting from the erosion of the volcanic islands. Present-day examples of volcanic island arcs are widely distributed in the western Pacific Ocean, the eastern Indian Ocean and in the Caribbean (see Fig. 6.1).

The continental-margin subduction zone

This type of subduction zone, which always precedes continent–continent collision, is represented in simplified form in Figure 6.16B. The descending oceanic plate causes melting to take place in the mantle wedge above the subduction zone, leading to the formation of an elevated volcanic arc on the continent of the overlying plate. Sediments and volcanic debris derived from the erosion of this volcanic massif are deposited in the offshore ocean trench. Present-day continental-margin subduction zones are situated along the Pacific margins of South and Central America.

Plate collision and mountain building

Where two continental plates collide, a zone of crustal overlap is created that leads to a great increase in the thickness of the continental crust (Fig. 6.16C). In some cases, crustal thicknesses of up to 80 km have been recorded (e.g. in the Alps), compared with an average 'normal' crustal thickness of around 33 km. Because continental crust is less dense and thus more buoyant than the underlying, denser, mantle, much of this extra crustal material becomes elevated to form mountain ranges with heights of up to 8 km above sea level. Due to the isostasy principle, these mountain ranges are supported by a much greater thickness of crustal material beneath, forming a kind of mountain 'root'.

The thickened crust of the collision zone consists of rocks that have been intensely deformed by the effects of the collision: in places, great thrust sheets will have formed due to the sliding of the upper plate over the lower. As the crustal material becomes depressed into the warmer regions at depth, the folding and shearing processes caused by the collision are enhanced by the heating of the rocks caused by rising magmas. The effects of heat and pressure at depth produce metamorphic changes in both sedimentary and igneous rocks.

One of the best present-day examples of an active collision zone is provided by the mountain ranges of southern Asia (e.g. the Himalayan, Pamir and Tien Shan ranges) where the Indian continent has collided with, and underthrust, Asia. The record of the gradual convergence of these two continents has been well documented from the ocean floor magnetic data. Unlike constructive boundaries and transform faults, where the zones of earthquakes or volcanic activity are relatively narrow, in continental collision zones they can be over a thousand kilometres wide. The surface separating the two opposing plates may descend for long distances beneath the surface of the upper plate and may also be folded and faulted in a complex manner.

Postscript

The development of the plate-tectonic theory in 1967–68 was one of the most significant breakthroughs in earth science history. Much research since then has focused on refining the model and applying it to older orogenic belts. Some of this work will be referred to in future chapters.

No single earth scientist can be credited with the invention of the plate-tectonic model. Most observers have named Tuzo Wilson as the true originator of the idea, and indeed he was the first to note the importance of transform faults and to invent the concept of the 'rigid plate'. However, Wilson was a student of Harry Hess and had the benefit of detailed knowledge of Hess's theory of sea-floor spreading, which has an equal claim to have been what gave rise to plate tectonics. Then there were the important contributions of the geophysicists who took Wilson's idea and developed it into what we now know as plate tectonics.

Among these scientists, it is impossible (and indeed would be invidious) to select the most significant. Three prominent geophysical establishments were involved: Lamont Geological Observatory at Columbia University in New York, Princeton University in New Jersey and Cambridge University in Britain. Lamont, under Jack Oliver, was responsible for the collation of the vast amount of seismic data needed to validate the theory, and it was here that the work of Isacks *et al.* and Le Pichon originated. Jason Morgan was at Princeton and studied under Hess. The Geophysics Department at Cambridge, under first Sir Edward Bullard and subsequently Drummond Matthews, had been studying global tectonics for many years. It was here that the palaeomagnetic work was carried out that 'proved' continental drift and sea-floor spreading, and where Dan McKenzie commenced his research under Matthews. It is also worth noting that both Hess and Wilson had visited Cambridge during these critical years (Wilson received a DSc from Cambridge to add to his earlier

BSc). So all these scientists were almost certainly familiar with each other's work, and thus the development of plate tectonics as a valid theory could be regarded almost as a collective endeavour.

7

Ophiolites: clues to the ocean crust and mantle

The presence of ophiolites has long been regarded as an important clue in the investigation of orogenic belts. Since they were recognized as representatives of oceanic crust and upper mantle in 1968 by Ian Gass, they have been widely used to mark the sites of former subduction zones and consequently to establish the boundaries between separate blocks (i.e. plates or 'terranes') of continental crust.

Historical background

The term 'ophiolite' was introduced in 1813 by French mineralogist Alexandre Brongniart to describe a group of green-coloured rocks, including serpentine and basalt, found together at several localities in the Alps. He redefined the term in 1821 to embrace a suite of igneous rocks including ultramafic varieties, gabbros, basalts and volcanics, and in 1927 the term was expanded by Gustav Steinmann to describe the triple assemblage of serpentine, pillow basalt and deep-marine chert, commonly found in the Alpine mountain belts, that subsequently became known as the Steinmann trinity (Fig. 7.1).

Steinmann believed that the ophiolite bodies found in the Alps and Apennines were intrusive into the deep-marine sediments with which they were associated, that the magma had probably been injected along elongate fractures and had subsequently differentiated into ultrabasic and gabbroic phases, to be followed by the basaltic lavas. The deep-ocean environment was indicated by the association with radiolarian chert. Until the advent of plate tectonics in the 1960s, the Steinmann assemblage was regarded as an essential component of 'eugeosynclines': those geosynclines thought to have been founded on oceanic crust at the margins of continents, in contrast with the 'miogeosynclines', which were viewed as deep basins developed entirely on continental crust.

The Alpine ophiolites first studied were typically affected by low-grade metamorphism characterized by the conversion of plagioclase to albite, pyroxene to chlorite and olivine to serpentine, which made their petrological study

Figure 7.1 The Steinmann Trinity: **A** Serpentinized peridotite; **B** basaltic pillow lavas; **C** bedded red chert. ©Shutterstock: A, by Sundry Photography, B, by corlaffra, C, by Weldon Schloneger.

more difficult than was subsequently the case with better preserved examples. However, their common association with copper ore, such as in the Cyprus example, was certainly an incentive for their investigation.

Throughout the 1950s and early 1960s, discussions on the origin of ophiolites centred on whether the ultramafic components were produced as a result of: a) the gravitational differentiation of ultrabasic material from an essentially basic magma, which had been intruded into the lower levels of eugeosynclines and/or erupted on the sea floor; or b) the tectonic emplacement of part of a mantle-derived peridotite layer. W.P. de Roever, in 1957, had suggested that the 'Alpine-type' peridotite massifs could be tectonic fragments of the peridotite layer, and in 1963, M. Vuagnat pointed out that the proportion of ultrabasic to basic material was far too high for the ultrabasic proportion to be a differentiation product of basic magma, and concluded that the peridotite massifs were the residual products of partial melting of the upper mantle.

As a result of the extensive research carried out on the ocean basins in the 1960s, it became clear that the serpentinization of peridotitic mantle material may play an important role in the structure of the mid-ocean ridges. In 1955, H.H. Hess had proposed that upwelling mantle currents would elevate the temperature beneath the ridges, resulting in the serpentinization of the mantle peridotite there, and that the Mohorovičić discontinuity (see chapter 4) may represent the boundary between serpentinized and unserpentinized peridotite, corresponding to the 500°C isotherm. The idea of sea-floor spreading, involving rising currents beneath ridges and descending currents at trenches, was now being widely discussed. Then, in 1963, F.J. Vine and D.H. Matthews published their ground-breaking analysis of the structure of ocean ridges, which was a major contributory factor leading to the plate-tectonic theory later in the decade.

However, in terms of the significance of ophiolites, the key step-change in understanding was made by Ian Gass in 1968, as a result of the investigation of the Troodos complex of Cyprus, by linking the Vine-Matthews model of ocean ridges with a detailed study of a well-exposed ophiolite occurrence.

The structure of the oceanic crust

Seismic structure

The investigation of the oceanic crust prior to the 1960s had been carried out mainly by means of seismic refraction and reflection methods. These had revealed a three-layered structure: an uppermost layer (layer 1), with an average thickness of 0.4 km, and consisting of sediments, which is thickest near the continental margin but decreases in thickness and locally disappears on the flanks of ocean ridges; a middle layer (layer 2) which has an average thickness of 1.5 km

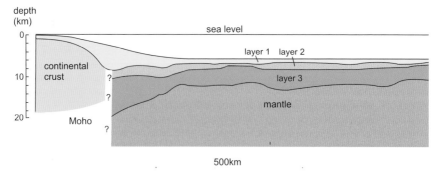

Figure 7.2 Oceanic crustal structure determined by seismic refraction in the Atlantic Ocean east of Argentina. After Ewing, 1965.

but is very variable, and a lowermost layer (layer 3), with an average thickness of 5 km, which is the most uniform both in thickness and wave velocity. This 3-layer crust, which has a total thickness of 6–7 km, rests on the Mohorovičić discontinuity, below which the seismic wave velocity increases sharply into a denser material assumed to represent the mantle (Fig. 7.2).

The seismic wave velocities of layer 2 were consistent with the physical properties of basalt, and the more recent work on the magnetic properties of the ocean floor described below confirmed this interpretation. Layer 3, which forms the bulk of the oceanic crust, had usually been interpreted as consolidated basic igneous rock such as gabbro, however, J.R. Cann in 1968 suggested that the material was more likely to be amphibolite, produced from the metamorphism of basalt in the presence of hydrous fluids at the ocean ridges.

The magnetic anomaly pattern

The regular pattern of magnetic anomalies on the ocean floor was first discovered and mapped off the west coast of North America by A.D. Raff and R.G. Mason in 1961, and subsequently discovered to be characteristic of the ocean floor generally. The pattern was interpreted in the celebrated paper by Vine and Matthews in 1963. They proposed that new basaltic crust forming along the ocean ridges becomes imprinted with the contemporary magnetic field, and that this changes periodically because the Earth's magnetic field reverses at irregular intervals, each of several hundred thousand to several million years long, after which the magnetic north and south poles are swapped. During each of these intervals a long strip of crust is created, parallel to the ridge axis, whose magnetic character differs from the previous one, and can be distinguished from it by remote measurement with a magnetometer. As new strips are created, older strips move away from the ridge axis. Thus each of the magnetic strips on the ocean floor represents a particular period of formation (Fig.

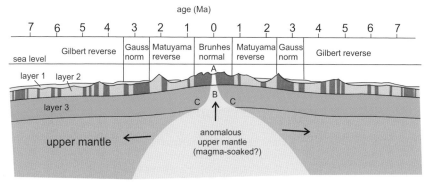

Figure 7.3 The ocean-floor spreading hypothesis as proposed by Vine & Matthews. The ocean crust is formed as the lithosphere on either side spreads laterally at about 1–6 cm per year. At A, layer 2 is formed by solidification of basic magma; at B there is a mush of ultrabasic upper-mantle rock, basic magma and water; at C, the Moho is frozen in at the 500°C isotherm as water and olivine react to form serpentine. In layer 2, normal magnetic polarity is shown in red, reverse in blue. After Bott, 1971.

7.3). The strip sequence was dated by comparing it with sequences of lava flows on land whose dates were known. This proved to be one of the most significant of the discoveries that led to the formulation of the plate-tectonic theory, as explained in the previous chapter.

The Troodos ophiolite complex

The contribution of Ian Gass

Ian Gass (1926–1992) is probably best known for having founded the extraordinarily successful Earth Science Department at the newly formed Open University at Milton Keynes in 1969. He was made a Fellow of the Royal Society in 1983. However, his scientific contribution mainly rests with his discovery of the significance of the Troodos ophiolite in Cyprus. Gass worked for the Cyprus Geological Survey from 1956 to 1960, and in 1963, he and D. Masson-Smith published an account of the geology of the complex together with the results of a gravity survey, which they interpreted as an indication that the floor of the complex was formed by continental rocks. They concluded that the complex might represent sub-Mohorovičić (i.e. upper mantle) peridotitic material which had partially fused to provide the volcanic material of the upper part of the complex, and that the whole complex had been underthrust by the leading edge of the African continent (Fig. 7.4).

Following on from the publication of the Vine-Matthews theory on ocean ridge formation, Gass's investigation of the ophiolite complex culminated in his 1968 paper *Is the Troodos Massif of Cyprus a fragment of Mesozoic ocean floor?*

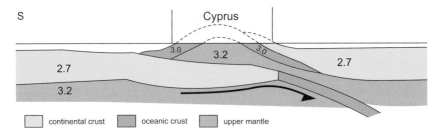

Figure 7.4 Interpretation by Gass and Masson-Smith (1963) based on a gravity survey showing the emplacement of the Troodos ophiolite complex as an obducted slab onto the African continent. The figures are the densities of: continental crust, oceanic crust and upper mantle used to interpret the gravity data. After Gass, 1980.

The Troodos ophiolite

Gass introduces his analysis of the Troodos ophiolite complex by referring to the recently published interpretation by Vine and Matthews that new ocean crust is created at the axes of mid-ocean ridges, and that oceanic magnetic anomaly patterns reflect strips of ocean floor created during periods of alternating normal and reversed magnetic polarity. He notes that few cases exist where this model can be tested on exposed ocean floor: oceanic islands are of little value because they represent the peaks of volcanoes whose chemistry has been affected by fractionation processes, and that, even in Iceland, which might be thought to be an obvious example of a mid-ocean ridge, the surface geology is dominated by effusive lavas. Moreover, the ophiolite masses found in the young orogenic belts, which have been identified as possible representatives of oceanic crustal layer 2, are usually metamorphosed, deformed and often allochthonous. In view of this, Gass suggested that the Cyprus ophiolite, because of its undeformed and (relatively) unaltered state, should be considered as an example of oceanic crust.

The Troodos massif is described as occupying an oval area of about 2300 km^2 of southern Cyprus (Figs 7.5, 7.6), and is surrounded by undeformed carbonate sediments of upper Cretaceous to Recent age. The complex is bounded in the north by the highly deformed Cretaceous to Miocene strata of the Kyrenia Range. The massif itself has a tripartite structure. The innermost part, the Troodos Plutonic Complex, is composed of basic and ultrabasic plutonic rocks, and is surrounded by a broad zone consisting of the Sheeted Dyke Complex dominated by basic dykes. The periphery of the massif is formed by the Troodos Pillow Lava Series, composed of basaltic pillow lavas and their related intrusive dykes.

Figure 7.5 The Troodos Mountains, Cyprus. © Shutterstock, by bensliman hassan.

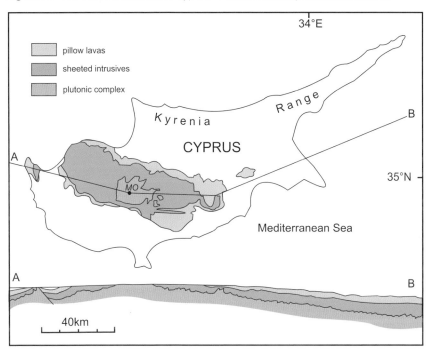

Figure 7.6 The Troodos Massif of Cyprus. MO, Mount Olympus. After Gass, 1968.

The Troodos Plutonic Complex

The rocks of the plutonic complex include dunites, peridotites, pyroxenites, gabbros, and marginal granophyres, becoming generally more silicic upwards and outwards. No floor to the complex has been recognized, and the surrounding host rocks are not metamorphosed. Gass concludes that these plutonic rocks belong to a differentiated ultrabasic body of batholithic dimensions: the proportion of ultrabasic material increases with depth, and the gravity data indicate that Cyprus is underlain by a vast mass of ultrabasic material.

The Sheeted Intrusive Complex

This consists almost entirely of N–S-trending basic dykes varying in width from 0.5 m to 3 m. In a cross-strike section of over 100 km, dykes occupy more than 90% of the outcrop, only occasionally separated by narrow screens of the host rock lavas. Gass concludes from the lack of evidence of deformation, and the uniformity of trend, that the dykes were injected during a period of E–W extension.

The Pillow Lava Series

The pillow lavas have an estimated thickness of around 1000 m and are of basaltic composition, although now largely affected by hydrothermal alteration. They are divided into two groups by a partial unconformity. There are numerous basaltic dykes that are regarded as feeders to the lavas, as they become less numerous upwards. The chemistry of both dykes and pillow lavas indicates a 'tholeiitic' composition, most closely resembling Hawaiian basalts. The copper-rich massive sulphide deposits that gave Cyprus its name (from the Latin *cuprum*) are contained within the upper parts of the lower pillow lavas.

Interpretation

Gass interprets the Troodos massif as:

> 'a Mesozoic volcanic pile of tholeiitic character that evolved in a subaqueous environment. The Plutonic Complex could be peridotitic material fused sufficiently so that gravitational differentiation could operate to form the layered sequence exhibited and to provide a liquid phase to be extruded to form the Pillow Lava Series.'

Gass concludes that the structure of the Sheeted Intrusive Complex meets the requirements of the Vine-Matthews model of a mid-ocean ridge, but states that, in view of the relatively short cross-strike section exposed in Cyprus, it was not possible to decide whether the Troodos complex actually lay across a ridge axis or on its flanks. The magnetic data show an anomaly pattern that is rather subdued but does exhibit a N–S orientation, parallel to the dykes of the intrusive complex and therefore to the presumed ridge axis. He points to the significance of the fact that the feeder dykes of the Pillow Lava Series, which are separated from the Sheeted Intrusive Complex by an unconformity, also possesses a N–S orientation, indicating that the focus of magmatic activity remained in the same area after a period of quiescence, suggesting:

> 'the Pillow Lava Series cannot be said to have issued from an axial zone unless the entire massif is axial ...'

and further that:

> 'If the Sheeted Intrusive Complex of the massif represents oceanic layer 2, then the Troodos Plutonic Complex may represent layer 3. The crudely stratiform arrangement ranging from central ultramafic varieties through gabbros to marginal silicic members is the type of structure that could be formed in areas of excessive heat flow, such as the crest of mid-ocean ridges, where the upper part of the mantle was fused sufficiently to provide a liquid phase to be injected as dykes [and] to be poured out as lava flows.'

The significance of this analysis by Gass is that, for the first time, we have a convincing example of the composition and structure of the oceanic crust and upper mantle at a possible ridge axis site and a test of the Vine-Matthews theoretical model.

Gass's paper stimulated an international flurry of activity: other ophiolite complexes were re-examined to establish whether the sequence recorded in Cyprus could be applied generally. Among the first to do so, in 1971, were W.R. Church and R.K. Stevens who described similar ophiolite sequences in the Newfoundland Appalachians, and in 1972 this period of renewed interest culminated in a Penrose Conference during which international agreement was reached in establishing a re-definition of the term 'ophiolite'.

The contribution of Moores and Vine

Eldridge Moores had been investigating North American ophiolite occurrences, and in 1969 re-investigated the Troodos Massif together with Fred Vine, which resulted in their paper entitled *The Troodos massif, Cyprus and other ophiolites as oceanic crust: evaluation and implications* published in 1971. This summarized the views reached at that time, and linked the formation of ophiolites firmly with the ocean-floor spreading process.

Moores and Vine improved the Gass model by pointing out the contrast between the ultramafic cumulates of the Plutonic Complex (i.e the dunites, pyroxenites etc) and the lower tectonized part (the harzburgite). They concluded that the dunite represented the base of the cumulate section produced by the differentiation of mantle-derived magmas, and that the foliated harzburgite beneath was deformed upper mantle (Fig. 7.7). The boundary between the two thus represented the base of the crust. Gass had thought that the

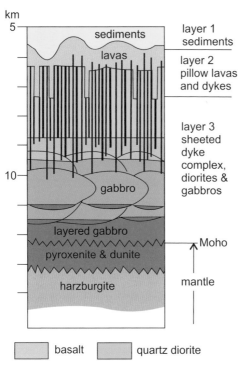

Figure 7.7 Cross-section of an oceanic spreading centre, based on the Troodos ophiolite. Note that the upper gabbros are shown as multiple magma chambers and the lower gabbros as layered. The upper peridotite layer represents cumulates. After Moores and Vine, 1972.

plutonic rocks had all belonged to a differentiated ultrabasic mass 'of batholithic dimensions'. The foliated harzburgite was believed to represent residual mantle material from which basic melt had been extracted and intruded into the crust.

The significance of this contribution was that a distinction was now being made between ultramafic igneous rocks that result from a magmatic crust-forming process and those that are part of the upper mantle and have never left it. Moreover, the base of the crust as defined by seismological criteria – the change from material with a density of c.2.9–3.0 to material with a density of c.3.3 (the 'seismological Moho') – can no longer be regarded as the true base of the crust (i.e. the 'petrological Moho'), which should be taken as the top of the upper mantle, and may be some distance beneath the seismological Moho.

The concept of the lithosphere

Several key concepts that affected views on ophiolites arose out of the debate during the period in 1968 and 1969 when the plate-tectonic model was being constructed. The idea of the crust being the mobile layer of the 'conveyor belt', visualized by Hess and others as the means of transporting the continents across the globe, was replaced by the new concept of the lithosphere – a layer based on its strength relative to the material beneath it.

The base of the lithosphere was defined as the boundary between 'strong' mantle material with a high seismic velocity and 'weak' mantle material with a relatively lower seismic velocity that occupied a zone known as the low velocity zone or LVZ. The LVZ had been discovered by Beno Gutenberg (see Gutenberg, 1959) and adapted for the purpose of the plate-tectonic model as the asthenosphere or weak layer over which the strong lithosphere could move. The asthenosphere was identified by geophysicists on the basis of its unusually low seismic shear-wave velocity and anomalously high electrical conductivity, thought to indicate a small proportion of melt (perhaps around 1%) possibly due to the presence of water. The top of the LVZ appeared to vary from c.50 km to c.100 km depth, and the base between 200 km and 250 km depth; the lithosphere thus varies in thickness – being thinnest near the ocean ridges and thickest near trenches.

The lithosphere–asthenosphere concept was significant for ophiolite interpretation: the tectonized peridotites with flow fabrics found beneath the ultramafic cumulates could be assigned to parts of the upper mantle that had belonged to the LVZ when at a spreading centre but had been added to the lithosphere as they cooled and moved away from the spreading centre. It would be expected that such rocks might include both fresh peridotitic material (from which no melt had been extracted) and depleted mantle rocks, which had partially melted to produce basalt magmas that had been injected into, and ultimately formed, the crust. Ophiolites should thus be seen as fragments of oceanic lithosphere that have become detached from the remainder of the lithosphere by a tectonic process, and the variety of ultramafic rock types that they exhibit can be explained by this model.

The structure of a typical spreading ridge

Figure 7.8 shows how the creation of new oceanic lithosphere may take place. Some of the new material is in the form of basaltic magma produced by melting of the upper mantle within the hot, low-density region beneath an ocean ridge. Part of this magma is injected into the crust in the form of intrusive bodies (i.e. dykes and sills), and some is poured out at the surface as lava flows. These rocks form the new oceanic crust; beneath it, new mantle lithosphere is formed by the addition of cooled peridotitic material transferred from the asthenosphere. The ridge as a whole may be over 1000 km wide and 2–3 km above the surrounding ocean floor, but the earthquake and volcanic activity is concentrated in a narrow central rift about 100 km across. The ridge is uplifted because of the presence of so much low-density material beneath it,

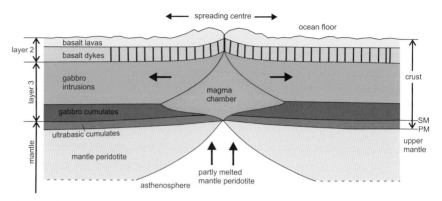

Figure 7.8 Petrological model of the oceanic crust – compare the seismological model of Figure 4.2. Note that the seismological Moho (SM) (the base of layer 3) is different from the petrological Moho (PM) which lies at the base of the ultrabasic cumulate layer. After Kusznir and Bott, 1976.

and this results in a process of stretching that enables the new material to rise into the crust. As this new material is added, earlier-formed lithosphere moves sideways, cools, and gradually sinks to the normal level of the deep ocean floor.

The problem of ophiolite emplacement

It was soon realized that oceanic lithosphere arriving at a subduction zone has little chance of ending up on the upper plate where it could be preserved, and various ingenious mechanisms were proposed to solve this problem. From the concept of ophiolites representing conditions at a mid-ocean ridge, as Gass had thought, opinion moved to consider ways of producing ophiolites on the upper plates of subduction zones – either as part of the 'fore-arc' region of a volcanic arc, or as part of a 'back-arc' spreading regime, or marginal basin as shown in Figure 7.9 (the formation of back-arc basins is the subject of chapter 9). Because these lie on the upper plate of a subduction zone, they have a better chance of being thrust onto an approaching continental margin and thereby preserved. These possibilities were discussed at the First Penrose Conference on ophiolites in 1969, and more extensively at the second, in 1998.

The first Penrose Conference on Ophiolites

Penrose conferences are held by the Geological Society of America to discuss topical subjects of international interest, and in 1969, the subject chosen was 'ophiolites'. The conference resulted in an agreed definition of an ophiolite, published in 1972, as follows:

> 'Ophiolite refers to a distinctive assemblage of mafic to ultramafic rocks. It should not be used as a rock name or as a lithologic unit in mapping. In a completely developed ophiolite, the rock types occur in the following sequence, starting from the bottom and working up:

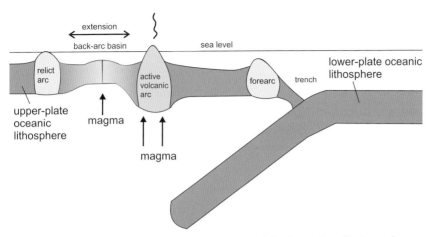

Figure 7.9 Cartoon cross-section of an intra-oceanic subduction regime. Sections of upper-plate oceanic lithosphere either from the back-arc basin or the fore-arc could be obducted as an ophiolite onto a continental margin approaching on the lower plate. Ophiolites formed in this way will display island-arc type features. Not to scale.

- ***ultramafic complex***, consisting of variable proportions of harzburgite, lherzolite and dunite, usually with a tectonic fabric (more or less serpentinized);
- ***gabbroic complex***, ordinarily with cumulus textures commonly containing cumulus peridotites and pyroxenites and usually less deformed than the ultramafic complex;
- ***mafic sheeted dyke complex***;
- ***mafic volcanic complex***, commonly pillowed;
- ***associated rock types***, including an overlying sedimentary section consisting of chert, minor shale and limestone; podiform bodies of chromite generally associated with dunite; and sodic felsic intrusive and extrusive rocks.

Faulted contacts between mappable units are common. Whole sections may be missing. An ophiolite may be incomplete, dismembered, or metamorphosed. Although ophiolite generally is interpreted to be oceanic crust and upper mantle, the use of the term should be independent of its supposed origin.' [Anonymous, 1972]

Post-Penrose-I developments

The problem of emplacement

By 1969, the plate-tectonic model had become widely accepted, and geologists had to fit ophiolites into a plate-tectonic context. It became obvious that the process of subduction would usually destroy normal oceanic crust (as part of the downgoing lithosphere), and that to ensure their preservation, ophiolite complexes would have to be formed on the upper plates of subduction zones. In 1971, Coleman introduced the term 'obduction' to describe the process of emplacement of an ophiolite onto continental crust, as had been suggested for the Troodos Massif by Gass and Masson-Smith. The process implies that the basal thrust of an ophiolite complex formed in this way should exhibit intense deformation and possibly incorporate slices of material derived from the downgoing slab. It appeared that for an ophiolite to be emplaced on continental crust, it would be necessary for a passive continental margin to arrive at a trench and underthrust a forearc region.

Accretionary orogenic belts such as those along the western Americas and in Japan are characterized by a wide variety of ophiolite occurrences, many of which are highly deformed and disrupted, of disparate ages and provenances, and represent separate blocks or slices of oceanic material brought to the surface as a result of a subduction-related process succeeded by the collision and accretion of the block to the continental margin. The term 'terrane' was introduced by Coney and others in 1980 to describe the separate blocks of displaced crust brought together in this way. Many of these either consist of ophiolites or are separated from each other by ophiolites.

Chemical composition

More data were now becoming available on the chemical composition of the products of present-day oceanic volcanicity in various tectonic environments: mid-ocean-ridge basalts, for example, were typically olivine-tholeiites, relatively rich in silica and poor in alkalis (Na_2O and K_2O): such basalts are referred to as N-MORB ('normal' mid-ocean-ridge basalts), whereas the volcanic rocks of intra-oceanic islands such as Hawaii are N-MORB in type in the earlier lava phases but alkali-basaltic, characterized by relatively higher alkalis and a lower silica content, in the later. Such basalts are distinguished as OIB (ocean-island basalts) or E-MORB (enriched MORB). Another category that was recognized was that of continental flood basalts, responsible for the huge outpourings of the Cretaceous Deccan traps, and more recently associated with present-day oceanic plateaux, and assigned to 'hot-spots' or 'plumes' (see chapter 10). These are now referred to as LIPs (large igneous provinces) and their basalts are dominantly tholeiitic.

In contrast, the volcanic products of subduction zones, the island-arc lavas, are characterized by a series of essentially silica-rich calc-alkaline types, ranging from tholeiitic basalts through andesites to dacites. These several categories are now much more easily distinguished by various trace-element ratios. The significance of the mineralogical and chemical composition of the various products of mantle melting will be explored in the discussion on hot-spots and plumes in chapter 10. However, the important point to make in relation to the study of ophiolites is that a conclusion as to the tectonic environment in which a particular ophiolite body originated could only be drawn from the consideration of the whole range of compositional types displayed.

In the case of the Troodos complex in particular, Akito Miyashiro argued in 1973 that, as one-third of the lower pillow lavas and the sheeted dykes are calc-alkaline rather than tholeiitic, they were more likely to have been produced in an island-arc setting with a relatively thin oceanic crust than at a mid-ocean ridge. Then Julian Pearce, in 1975, suggested that Troodos might have formed in a marginal or back-arc basin during the evolution of an island arc, as indicated in Figure 7.9. However, the lack of a volcaniclastic component in the Troodos complex, which might have been expected near an island arc, was puzzling. In the same year, Miyashiro proposed a new threefold classification of ophiolites:

1 **island-arc type**, which included both tholeiitic and calc-alkaline volcanics;

2 **tholeiitic type**, which could be of either island-arc or mid-ocean ridge origin;

3 **tholeiitic plus alkaline type**, which originated in an extensional rift setting at or near an intra-oceanic volcano.

The Semail ophiolite

This ophiolite complex is situated in the Oman Mountains, extending for about 450 km on the southern side of the Gulf of Oman. It is probably the largest ($c.30,000\,\text{km}^3$) and best exposed ophiolite yet found, and was mapped in the 1960s by the Shell Oil Company (e.g. Glennie *et al.* 1974). Figure 7.10 shows

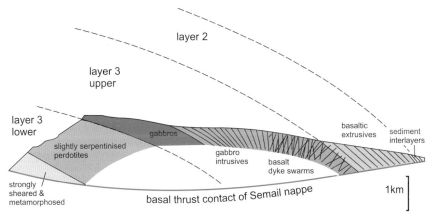

Figure 7.10 The Semail ophiolite complex, Oman. The peridotites and gabbros include both massive and cumulate types. The lowermost rocks include mylonitized high-grade metamorphic units. After Glennie *et al.*, 1974.

a simplified section through the ophiolite, from which it is evident that the complex represents a complete c.12 km-thick sequence through the oceanic crust. Unlike the Troodos complex, the base of the Semail ophiolite is clearly visible, and rests on a thrust plane on which the ophiolite has been transported onto the Mesozoic platform sediments of the Arabian shield.

Classification of ophiolites

In addition to the mid-ocean ridge environment originally proposed for Cyprus and other similar ophiolite occurrences, it had become clear that ophiolites represented a range of compositions, structures and environments. It was concluded that, in general, ophiolites represent ocean crust and mantle formed at ocean spreading centres, but that these centres could be situated either at mid-ocean ridges, in pull-apart intra-arc basins, or at extensional zones in the centres of fore-arcs during the initial development of island arcs. In terms of internal structure, some ophiolites reveal the complete (Penrose) sequence, while others lack a sheeted dyke element, and are interpreted as the result of developing in a magma-starved environment such as that of modern slow-spreading ridges.

In 1982, Moores classified ophiolites into Tethyan and Cordilleran, based on whether they rested on a passive continental margin (as most of the southern Alpine examples appeared to do) and had therefore been obducted, or alternatively had formed as part of an island arc assemblage and had been incorporated into a convergent active continental margin, as in the case of the Cordilleran examples. The evidence in some ophiolites of metamorphic retrogression in the lower part of the sequence suggested that the material at the base had been cooled by the descending cold slab, reinforcing their status as essentially upper-plate phenomena.

Another conclusion that was reached after studying present-day examples of small oceanic basins around the western Pacific rim was that there are many such basins that are of relatively short duration (less than 10 Ma) and are the product of young, relatively hot lithosphere. This may explain the more MORB-like chemistry and the lack of evidence of volcanic arcs found in many ophiolite occurrences.

The second Penrose Conference on ophiolites

A second Penrose meeting on ophiolites was held in 1998 to re-evaluate models of ophiolite genesis. A new classification of ophiolite types was produced to account for the large variety of types found during the intervening period since Penrose I. These were summarized by Yildirem Dilek in 2003.

1 *Ligurian type*. These include serpentinized peridotites intruded by gabbros, and local dykes, and overlain by pillow lavas, but lack a sheeted

dyke complex. Contacts between the various units are either intrusive, tectonic or stratigraphic, and the mantle rocks may be significantly older than the crustal component. The peridotites are mainly lherzolites and exhibit high-temperature fabrics, while the lavas have a mid-ocean-ridge (tholeiitic) chemistry. These ophiolites are considered to represent an early stage of the opening of an ocean basin following continental rifting.

2   **Mediterranean type**. These correspond to the Troodos example, with the complete Penrose sequence. They show extensional structures within the sheeted dyke complex and dykes are commonly feeders to the lavas. The composition of the ultramafic rocks is mainly harzburgite-lherzolite or harzburgite, reflecting the melting of depleted mantle, and are considered to be parental to tholeiitic to calc-alkaline magmas of island-arc type. These ophiolites are interpreted as having either a fore-arc or back-arc origin related to an intra-oceanic subduction zone, with their emplacement controlled by the relative movements of small plates such as those currently found in the western Pacific.

3   **Sierran type**. These are found typically around the Pacific rim, for example in Japan and California. Their volcanic rocks range from basalts to rhyolites and include volcaniclastics. They exhibit a history of multiple tectonic episodes, including rifting and strike-slip faulting and belong to a regime of oblique convergence. They are interpreted as the result of the accretion of an intra-oceanic island-arc onto a continent.

4   **Chilean type.** These comprise the whole range of Penrose units, including pilllow lavas, sheeted dykes and massive gabbros, with intrusive contacts. The igneous rocks show mid-ocean-ridge, tholeiitic, chemistry. They are interpreted as relatively autochthonous oceanic crust produced in a rift-generated back-arc basin developed within continental crust.

5   **Macquarie type**. This type is represented by a (possibly) unique occurrence on Macquarie Island, southwest of New Zealand. It includes the full range of Penrose units, and exhibits typical to enriched (more alkalic) mid-ocean-ridge chemistry. It was formed as the result of the uplift of a short segment of ocean ridge, due to oblique convergent movement along the junction between two oceanic plates.

6   **Caribbean type**. These are highly heterogeneous; they contain most Penrose units except for the sheeted dykes, which are usually missing. The chemistry is variable, ranging from normal to enriched mid-ocean-ridge type. They contain abnormally thick crustal sections, and are interpreted as the product of an oceanic plateau, part of what has become known as a 'large igneous province' or LIP. Tectonically emplaced fragments of Caribbean type ophiolites are thought to be common within orogenic belts.

7 *Franciscan type.* These are spatially associated with accretionary complexes along active continental margins (such as the western USA) and
are commonly intercalated with 'mélanges' and high-pressure metamorphic assemblages characteristic of subduction zones. They exhibit diverse
lithologies, metamorphic types and chemical affinities, and the various
units may have no apparent links between them. They represent tectonically imbricated slices of oceanic rocks scraped together from subducted
slabs.

Postscript

The significance of the work of Gass and his colleagues on the Troodos complex
was twofold. In the first place, it stimulated further detailed investigation of
the Cyprus ophiolite, and later of the well-exposed Semail ophiolite in Oman,
resulting for the first time in a believable petrological model of an oceanic
spreading centre, and enabled the various ultrabasic and basic igneous rocks
found in such complexes to be linked together into a petrogenetic model, as
shown in Figure 7.8. This in turn fed into the investigation of how mantle
convection works, which will be discussed in chapter 10 on mantle plumes.
Secondly, when linked with the plate-tectonic model, the Troodos investigation led to the recognition that most, if not all, ophiolite occurrences could be
interpreted as sutures between different continental plates or terranes – representing the former existence of oceanic basins that had been obducted onto an
adjacent continental block. The expanding information about the present-day
oceanic crustal structure of the western Pacific rim and Indonesia had revealed
the presence of many relatively short-lived back-arc or marginal basins, and it
was realized that the capture of spreading centres in these kinds of environment was the likely explanation for many of the ophiolite occurrences within
complex orogenic belts.

8

Fault system kinematics

Historical background

Faults have been known since the earliest days of the mining industry, as breaks in the continuity of individual beds or formations, but they have been given various names because each mining district tended to develop its own specialized terminology. The term 'fault' did not become generally used until the twentieth century. The eighteenth-century field scientists such as James Hutton, William Smith and Charles Lyell say very little about faults, although they clearly recognized that they exist. In contrast, the theory of fracturing of materials had been formulated in the mid-eighteenth century. Eduard Suess is credited with the first attempt at classification of faults in 1904 when he divided faults into 'overthrusts' (*Wechsel*), 'wrench faults' (*Blätter*) and 'normal faults' (*radial faults*).

Some of the nineteenth-century geologists who investigated the great thrust systems of the Alps and the Scottish Highlands were concerned with how these systems developed. For example, John Horne, in chapter 32 of the famous 1907 NW Highlands Memoir, discusses how the thrusts and related folds of the Moine thrust belt could have originated as the result of a sequence of related movements.

However, for about three decades during the middle of the twentieth century, most students of structural geology (at least in Britain) based their knowledge of faulting theory on the book by E.M. Anderson entitled *The dynamics of faulting and dyke formation with applications to Britain*, published in 1942. As the title of his book indicates, Anderson's approach was to view the study and classification of faults entirely as a problem of dynamics: he maintained that faults are a response to a particular stress regime, which could be reconstructed by observing the orientations and sense of shear of pairs of faults, as illustrated in Figure 8.1. A similar approach is taken by Neville Price in his 1966 textbook *Fault and joint development in brittle and semi-brittle rocks*, albeit in much greater depth. Both these authors treat faulting essentially as a problem of brittle failure in response to stress.

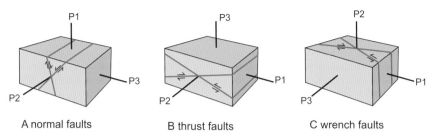

Figure 8.1 Orientation of shear fault sets in relation to principal stress axes: P1>P2>P3 as proposed by Anderson, 1942.

Structural geologists in Britain during the 1960s and 1970s were focused much more on the processes governing folding and fabric formation than on faulting. In one of the most influential textbooks of that period, *Folding and fracturing of rocks* by John Ramsay, published in 1967, faulting is treated either in terms of brittle failure, or as a secondary response to folding.

Meanwhile, across the Atlantic, a quite different approach was taken by structural geologists working in the Appalachian and Rocky Mountains fold-thrust belts. There, the emphasis was on attempting to reconstruct the complex geometry of these belts, and to understand the movements that had given rise to the structures. In other words, they adopted a 'kinematic' (movement-based), rather than a 'dynamic' (stress-based) approach. This was especially true of the petroleum exploration geologists, whose decisions about sub-surface geometry would have critical economic implications. Two geologists working in the Canadian Rockies were particularly influential during this period – Albert Bally, of Shell Canada, and Clint Dahlstrom of Chevron Standard, and further south, in the USA, Vinton Gwinn of Mobil Oil was examining the tectonics of the Central Appalachian fold-thrust belt. At about the same time, in mainland Europe, the investigation of the Jura fold belt of the Western Alps was being undertaken by Hans Laubscher.

Thin-skinned tectonics

The 'thin-skinned' model of fold-thrust belts refers to a style of deformation where the folding and thrusting are confined to an upper layer of the crust, and separated by a detachment (or 'décollement') horizon from a basement that is left relatively unaffected. The thin-skinned model had been discussed for many years, especially in respect of the Jura, the outer zones of the Appalachians and the Rockies. The Jura fold-thrust belt, in particular, had become almost a 'type' example of the thin-skinned model, and was the subject of two important papers by Hans Laubscher, in 1962 and 1965, in which he suggested methods of reconstructing the geometry of the fold-thrust complex and of determining the amount of shortening across the belt.

No general agreement had been achieved in respect of the Appalachian and Rockies belts, however, and the competing 'thick-skinned' and 'thin-skinned' models were much debated. It was mainly because of the subsurface evidence from boreholes and reflection seismic data obtained by the oil industry that the thin-skinned model for these belts was finally proved beyond doubt.

Two key papers emerged as a result of this North American work during the period 1964–1966: by Vinton Gwinn, in 1964, on *Thin-skinned tectonics in the Plateau and Northwestern Valley and Ridge Provinces of the Central Appalachians*, and the ground-breaking paper by Albert Bally (jointly with P.L. Gordy and G.A. Stewart), in 1966, on *Structure, seismic data and orogenic evolution of southern Canadian Rocky Mountains*. Three years later, Clint Dahlstrom published his paper on *Balanced cross-sections* which formalized the geometric methods by which the restoration of thin-skinned tectonic cross-sections could be achieved.

No special significance should be attached to the order in which these papers appeared in press, as the oil company geologists had been working for many years on their respective problems but had been restricted in publishing commercially sensitive material for years after the material in question had been obtained. Neither Bally nor Dahlstrom make reference to Gwinn's or Laubscher's work.

The contribution of Vinton Gwinn

Gwinn's study area was the Appalachian Plateau and Northwest Valley and Ridge provinces of the Central Appalachians, which he began investigating while working for Mobil Oil. Although thrusting was known to have played an important role in the deformation of the Southern Appalachians, further north, in the Central Appalachians, only folds are apparent at the surface. Gwinn's work was important in demonstrating that these folds were generated by thrust faulting at depth, and that the fundamental kinematic regime here was the same as that of the Southern Appalachians, but was represented at a higher crustal level. The thin-skinned kinematic model thus applied equally to both central and southern sectors of the fold belt.

Gwinn noted that deep wells in this part of the Appalachians penetrate major overthrusts that form part of a basal thrust fault (or 'sole thrust') system, and that most of the anticlines in the western part of the fold belt are cut by steep faults that curve downwards to connect with more gently dipping thrusts (Fig. 8.2). The folds were therefore a 'passive' response to step-like faults branching up from the basal sole thrust. This sole thrust initially follows a detachment horizon in mid-Cambrian shales, but moves progressively upwards into higher-level detachments in the Upper Ordovician and Upper Silurian respectively.

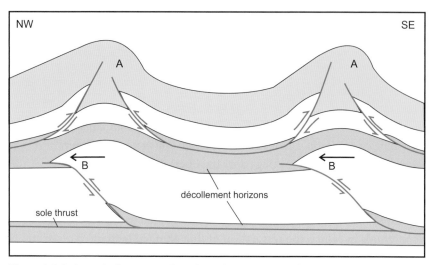

Figure 8.2 Schematic diagram to show Gwinn's explanation of how the anticlinal folds are produced. Those at B are formed by upthrusts from the basal sole thrust situated within the lower detachment horizon, while the folds at A are tightened by subsidiary thrusting from an upper detachment horizon. The asymmetric nature of the upthrusting at B causes the folds at A also to be asymmetric, with steeper SE limbs. Detachment zones in green. After Gwinn, 1964.

Gwinn's model explained an otherwise puzzling fold style where the few widely spaced anticlines were separated by broad, flat, synclinal areas. The 'active' parts of the fold system were the anticlines, whereas the synclines were merely those parts of the stratigraphic sequence overlying the flat thrust zones and thus unaffected by the up-thrusts.

Another prominent feature of the Central Appalachian fold belt is the periclinal nature of the anticlinal folds – 'periclines' are folds whose amplitude decreases to zero at both ends, producing an elongated oval outcrop pattern (Fig. 8.3). These could be explained by along-strike variations in the amount of displacement on the related thrust. The region is traversed by a number of NW–SE-oriented lineaments, some of which have been attributed to transcurrent faults; these may define the boundary between two thrust sheets advancing at different rates, or where one side cuts up to a higher level. The transcurrent faults need not penetrate the surface but could define the ends of the periclinal folds.

Gwinn's interpretation of how the fold belt evolved is shown in a cross-section (Fig. 8.4).

There are two important conclusions:

1 'Deformation of the folded Central Appalachians does not persist to the basement; bore holes penetrate major sole thrusts and branching splay thrusts in complexly deformed subsurface portions of Appalachian anti-clines which are unfaulted concentric folds at the surface.'

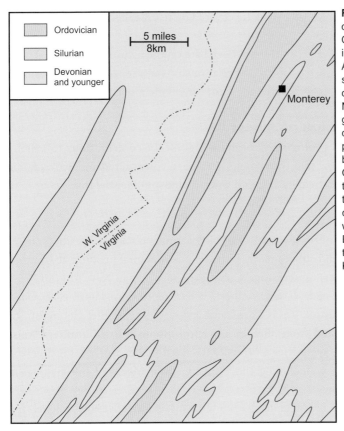

Ordovician	
Silurian	
Devonian and younger	

5 miles
8km

Monterey

W. Virginia
Virginia

Figure 8.3 Map of the Monterey Quadrangle in the Central Appalachians showing the style of the folding. Note the elongated oval shapes of the anticlinal periclines (in pink) bringing the older Ordovician strata to the surface, and the synclinal periclines (in green) with younger Devonian rocks in their cores. After King, 1951.

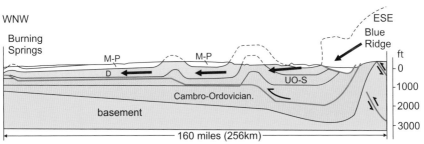

WNW

ESE

Burning Springs

Blue Ridge

M-P

M-P

D

UO-S

Cambro-Ordovician.

basement

ft

0

1000

2000

3000

160 miles (256km)

Figure 8.4 Schematic cross-section of the Central Appalachians showing the near-horizontal detachment zone (red), initially within the Cambro-Ordovician, climbing up into the base of the Upper Ordovician, then into the top of the Silurian. The source of the NW-wards movements is the crystalline massif of the Blue Ridge Province, which is interpreted as having been thrust upwards and emplaced on a higher thrust. UOS, Upper Ordovician-Silurian; D, Devonian; M-P, Mississipian-Pensylvanian. Thrusts in red. After Gwinn, 1964.

2 'The tectonic style of the Central Appalachians is entirely analogous to that of the Southern Appalachians' (the only difference being that) 'erosion has progressed far enough to expose the forward toes of the thrust faults of the Southern Appalachian thrust faults whereas those in

the Central Appalachians are covered by an unfaulted blanket of younger sediments.'

Gwinn's major contribution to the debate on the nature of fold-thrust belts is his detailed reconstruction of the relationship between folding and faulting, seeing both as an expression of the same kinematic movement.

The contribution of Albert Bally

Although several geologists have made significant contributions to solving the problem of how fold-thrust belts have been constructed and evolved, it is Albert Bally who is widely credited with the most influential role.

Albert Bally (1925–2019)

Bally was born in The Hague, in the Netherlands, obtained his doctorate at Zurich in 1952, and subsequently gained a post-doctoral position at Lamont-Doherty Geological Observatory, Columbia University, in New York. In 1954, he joined Shell Canada, based in Alberta, where he remained until 1966, when he moved to Houston, Texas as research manager, then Chief Geologist, for Shell US. He retired from Shell in 1981 to become Harry Carothers Weiss Professor of Geology at Rice University in Houston. His work was recognized with several awards, including the William Smith Medal of the Geological Society of London.

Bally was able to integrate the impressive seismic reflection and well log data accumulated by Shell for the purpose of petroleum exploration, along with surface mapping, to produce usable geometric reconstructions of the Rockies fold-thrust belt. His most influential paper, as mentioned above, was published in 1966, jointly with P.L. Gordy and G.A. Stewart.

One of the most important results of this work was the conclusive demonstration that the Precambrian basement extended westwards at a gentle inclination beneath the fold-thrust belt, proving that it was 'thin-skinned'– that is, that the folding and thrusting were essentially confined to the Phanerozoic cover and that the basement was relatively unaffected. Their structural cross-section was the first detailed interpretation of how the surface structures had been produced by tectonic movements at depth, and provided a believable model of the fold belt restored to its pre-deformation state.

The Southern Canadian Rocky Mountains

The study by Bally and his colleagues at Shell Canada was undertaken in a section of the Southern Canadian Rocky Mountain fold belt straddling the British Columbia–Alberta border and extending from the North Saskatchewan River to the border with the USA – a distance of about 260 miles (c.418 km).

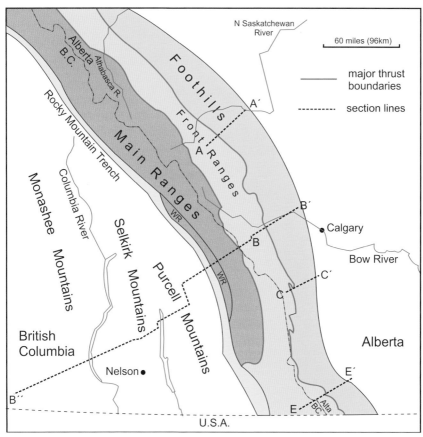

Figure 8.5 Map of the Southern Rocky Mountains showing the geological provinces. The provinces covered by the detailed geological and seismic sections are shown in colour. See Figure 8.6 for sections A–A′ and B–B′. The SW extension of B–B′ to B′′ refers to the authors' restored cross-section through to the western edge of the Monashee Mountains. WR, Western Ranges. Based on Bally *et al.*, 1966.

This particular part of the fold belt, which here is between 50 km and 60 km wide, is known as the Eastern Cordillera (Fig. 8.5). Seismic and gravity surveys, available from the 1940s, had led to the exploitation of the first gas fields, and during the 1950s and 1960s, extensive regional seismic reflection surveys were undertaken in the search for further petroleum fields. The seismic data on which this study was based was finally released by Shell to enable their paper to be published.

The Eastern Cordillera consists of folded and thrust Palaeozoic to Mesozoic sedimentary strata lying above a largely undisturbed Precambrian basement, which corresponds to the western continuation of the Canadian Shield. Between the deformed cover and the basement is the décollement, or detachment, zone, which consists of a weak layer located (in this area) within Cambrian shales.

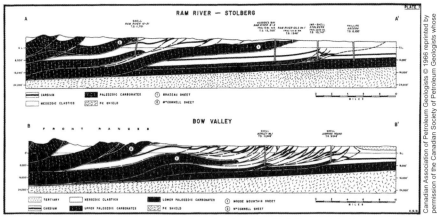

Canadian Association of Petroleum Geologists © 1966 reprinted by permission of the Canadian Society of Petroleum Geologists whose permission is required for further use.

Figure 8.6 Detailed sections along the lines A–A′ and B–B′ of Figure 8.1. Note position of well sections. From Bally *et al.*, 1966, with permission.

The Eastern Cordillera is further subdivided into an eastern sector, known as the 'Foothills', and a western, known as the 'Front Ranges'. The structure of the Foothills is dominated by large, generally flat-lying, thrust sheets consisting of Palaeozoic carbonates, overlain by steeper thrust slices of Mesozoic clastic material. The Front Ranges consist of stacked thrust slices composed of thick Palaeozoic carbonate units (Fig. 8.6). In the southern part of the cordillera, the Front Ranges also include thick thrust sheets of Late Precambrian clastic sediments known as the 'Beltian'.

West of the Eastern Cordillera lie the high main ranges of the Rocky Mountains: from east to west, the Eastern Main ranges, the Western Ranges, the Purcell Mountains, and the Selkirk-Monashee Mountains, which extend to the west coast. The Western Ranges are separated from the Purcell Mountains by the Rocky Mountain Trench, which is interpreted as a later extensional structure. Only the Eastern Cordillera is covered by the detailed sections controlled by the seismic and well data, but the authors' reconstruction extends west to include the Purcell Mountains.

The authors attempted to answer the following questions:

1   Does folding precede thrusting?
2   Are folded faults formed during or after the faulting process?
3   Does deformation proceed from west to east or from east to west?
4   Does normal faulting precede or follow thrust faulting?

Conclusions

The authors concluded that the deformation proceeded from west to east, starting during the Mesozoic and reaching the Foothills between the Palae-ocene and the Oligocene, during which time, they estimate a shortening of

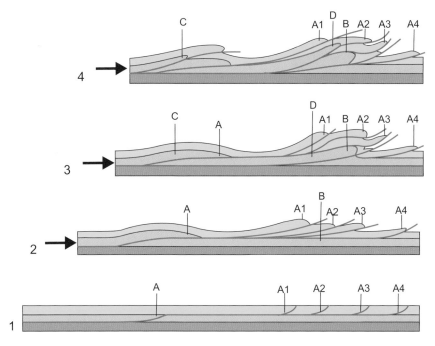

Figure 8.7 Sequence of thrusting in the Foothills Province during the main orogenic movements. Phase 1: initiation of step-like faults at A and A1–A4 from a sole thrust originating along a décollement horizon at the top of the basement (brown layer), rising to another on the top of the Palaeozoic (green layer). Phase 2: thrust sheet A is formed within the Palaeozoic, resulting in imbricate fan A1–A4 within the Mesozoic (yellow layer). Phase 3: a new deeper-level thrust sheet B forms within the Palaeozoic by the sole thrust proceeding further along the top of the basement; this results in the folding of the higher thrust sheets A2–A4. Phase 4: new thrusts develops within the Palaeozoic at C and D resulting in further folding of the upper thrust sheets A1 and A2. Based on Bally *et al.*, 1966.

about 25 miles (40 km) had been accomplished, representing around 40% of the original width of the belt (Fig. 8.7). They suggest further that the eastward progression was accomplished by means of a sequence of step-like faults which cut up from a basal thrust, in each case involving folding at the leading edge of the thrust sheet. As each new thrust developed, older thrust sheets were carried passively above it. This process necessarily leads to the folding of the previously formed thrust:

> 'In conclusion, we interpret that deformation of the Eastern Cordillera proceeded from west to east, that is, that the higher and more westerly thrust sheets were formed prior to the lower and more easterly elements.'

As the uplifted region produced during the early thrusting proceeded from west to east, the authors deduced that the foredeep basin that developed at its leading edge would also have migrated eastwards, commencing in the Upper

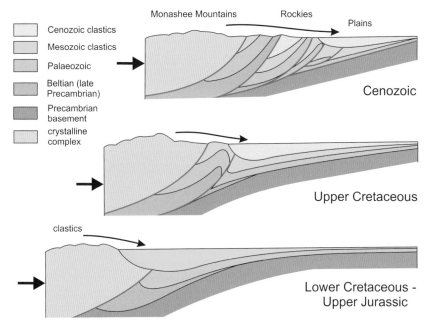

Figure 8.8 The migrating foredeep basin. As the fold-thrust belt develops, the mountain front moves eastwards and the depositional basin moves eastwards in front of it. After Bally *et al.*, 1966.

Jurassic and ending in the Cenozoic with the existing undeformed basin of the Western Plains, east of the deformation front (Fig. 8.8).

The final phase of deformation was caused by the isostatic uplift of the greater part of western Canada, and led to the formation of the present mountain topography. This process continued from the Oligocene through to the present day. During the preceding tectonic phases, the underlying basement was progressively tilted westwards; however, the re-adjustment brought about by the final uplift only reduced the degree of tilt but did not remove it, as can be seen in the cross-section.

The publication of this work made a huge impact on the world of structural geology, and led to the re-examination of many other fold-thrust belts using the same methods. The most significant aspects can be summarized as follows.

1 A highly deformed fold-thrust belt can be unravelled using a step-by-step process according to simple geometric rules, and generally follows a simple outward progression towards an undeformed foreland.

2 Older, higher thrust sheets are carried forward on top of younger, lower ones and are deformed with them.

3 As the mountain front progresses outwards with the deformation, the foredeep basin developed at its leading edge progresses outwards with it.

Three years later, an account of the geometric processes necessary to construct a believable restoration of a fold-thrust section was published in another influential paper by Clint Dahlstrom.

Balanced sections

The contribution of Clint Dahlstrom

C.D.A. Dahlstrom (1926–2015) graduated at the University of Saskatchewan in 1952, after which he proceeded to Princeton where he gained his PhD. In 1955 he joined Chevron Standard in Calgary where he remained until 1970, when he transferred to San Francisco. He retired in 1990. Dahlstrom was one of a group of structural geologists (including Bally), working for the various oil companies in the Canadian Rockies, who were collectively responsible for major advances in the understanding of fold-thrust belts. Dahlstrom's most important contribution was his paper *Balanced cross-sections* published in 1969.

What is a balanced section?

The purpose of a 'balanced section' is to check a structural cross-section for geometric acceptability. The idea behind it is that if a deformed section could be 'flattened out' in such a way as to restore it to its original undeformed state, the section is geometrically possible (although not necessarily correct). Dahlstrom stressed that his methods could only be applied in a restricted type of geological environment – the marginal parts of an orogenic belt such as that of the Foothills Province of the Canadian Rockies – and warned against applying them in other situations without modification. The method is illustrated by following the steps in Figure 8.7 in reverse: starting with 4 and proceeding to the restored balanced section 1. Bally's four conditions were as follows:

1   There must be a basement extending beneath the fold-thrust belt but which is itself unaffected by the folding and thrusting.
2   The rock sequence of the foreland must be essentially undeformed.
3   The deformation of the fold-thrust belt must have occurred long after most of the strata in question had been laid down (i.e. there will not have been any significant volume change resulting from the deformation).
4   The strata of the deformed belt must not have experienced any thinning or thickening – i.e. the folds must be 'concentric'* with no component of flow.

It is assumed that the cross-sections are taken parallel to the direction of tectonic transport which, in the case of the Foothills Province, is perpendicular to the strike. The reason why these conditions are important is that, in order to reconstruct the original geometry of a set of beds, it is necessary to assume that

*Author's note: it is only necessary that the folds be 'parallel', and this only need apply to those beds used in the calculation.

the beds have not changed in length as a result of the deformation – a condition met by parallel folds but not by folds of a more 'similar' type that exhibit a degree of flow oblique to the bedding. In other words, it must be assumed that the surface area of the deformed beds in the plane of the cross-section has not changed. This leads to a simple test of any reconstruction: that the restored bed lengths at several different horizons should be equal, as measured between two reference lines (subsequently to be known as 'pin lines'). These must be located along the axial planes of major synclines or anticlines or in other regions of no inter-bed slip, such as the undeformed foreland. If the bed lengths do not correspond, there must be an undetected discontinuity, such as a décollement horizon or thrust, separating the non-matching beds.

A cross-section that has been tested in this way, where all the measured bed lengths match, is a balanced section. Dahlstrom emphasized that such a section is not necessarily correct, but the fact that it balances reduces the chances of it being incorrect. Previous cross-sections of the Rockies fold belt included faults that ended downwards, without any explanation of how they were connected kinematically. In contrast, in the sections produced by Bally and his colleagues, all the faults are ultimately linked to displacements along the basal sole thrust and formed a 'closed system' in which all the displacement is eventually, in Dahlstrom's words: 'hustled westwards through the only available exit'.

As well as all the bed lengths matching in a balanced section, the displacements on individual thrusts must also be consistent. Thus where displacement appears to change along the course of a fault, this has to be accounted for either by substituting some fold shortening for fault displacement or by distributing the fault displacement along one or more splay faults, as shown in Figure 8.9.

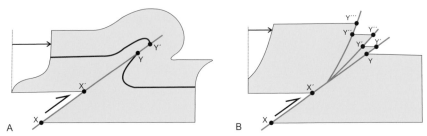

Figure 8.9 Accommodation of thrust displacement by folding (**A**) or imbrication (**B**). In **A**, the displacement X–X´ is partly taken up by displacement Y–Y´ on the thrust and partly by the folding. In **B**, displacement on the thrust X–X´ is distributed among the imbricate thrusts Y–Y´. Y´–Y´´ and Y´´–Y´´´. A consequence of this transfer of displacement is that deformation must take place within the yellow layer (e.g. by bedding-plane slip) shown by bending of the edges of the layer. Based on Dahlstrom, 1969.

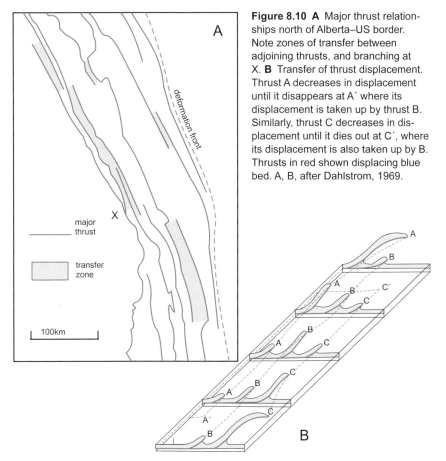

Figure 8.10 A Major thrust relationships north of Alberta–US border. Note zones of transfer between adjoining thrusts, and branching at X. **B** Transfer of thrust displacement. Thrust A decreases in displacement until it disappears at A′ where its displacement is taken up by thrust B. Similarly, thrust C decreases in displacement until it dies out at C′, where its displacement is also taken up by B. Thrusts in red shown displacing blue bed. A, B, after Dahlstrom, 1969.

Lateral changes in displacement

The amount of shortening in particular beds ought to be more or less the same in adjoining cross-sections. Gradual lateral changes are to be expected, but any sudden change must be due to the presence of an unsuspected fault between the two sections. In the case of the Foothills Province, the total shortening calculated for the different sections across the belt corresponds closely, so that where the displacement on one thrust decreases it must be taken up by increasing displacement on an adjoining thrust in order to maintain the constant overall shortening, as shown in Figures 8.10A and B. The areas of overlap between the pairs of thrusts correspond to 'transfer zones' in which the thrusts involved must be rooted in a common sole thrust for the system to be geometrically possible.

Conclusion

Dahlstrom ends with the following statement:

'In geology applied to oil and mining exploration or to engineering projects, cross-sections are used to convey predictions as to rock behaviour and they must be conceptually and, more importantly, geometrically correct. In these areas it is important to the geologist and for his client that there be some way of checking cross-section interpretation prior to drilling. Although a cross-section which passes the geometric tests is not necessarily correct, if it does not, it cannot possibly be correct.'

The methods that Dahlstrom sets out in his paper had been used by a number of geologists previously and had become routine practice among the petroleum exploration geologists working in the Canadian Rockies. However, Dahlstrom was the first to explain the process and lay down the rules in such a way that they could be universally applied. The publication of his paper sparked a major burst of activity among structural geologists worldwide, who rushed to re-interpret their local fold-thrust belts using his methods. Among the first to do so were David Elliott and Michael Johnson, who applied them to re-interpret the geometry of the Moine thrust zone in the Caledonian oro-genic belt of NW Scotland.

Later developments

The contribution of David Elliott

Although it was Bally and Dahlstrom whose work introduced the geometrical methods of reconstructing the kinematic processes of thin-skinned fold-thrust belts, the geologist who, probably more than any other, was responsible for spreading these ideas to Britain and the wider world was David Elliott.

David Elliott (1938–1982), after graduating at McGill University, Mon-treal, had gone to Glasgow to obtain his PhD, after which he proceeded to Imperial College, London, to work with John Ramsay. Elliott had become familiar, while at McGill, with the Canadian methods of structural analysis as a result of summer fieldwork in the Canadian Arctic, and during his time at Imperial College, his ideas influenced a number of other British structural geologists, including Michael Johnson and Michael Coward, both of whom applied the methods of section balancing to the investigation of the Moine thrust zone. The first study of the Moine thrust zone using the new methods of analysis was published by Elliott and Johnson in 1980 (Fig. 8.11).

In 1965, Elliott left London to take up the post of Associate Professor of Structural Geology at Johns Hopkins University, Baltimore, where he carried out work on various thrust belts, including the Appalachians and the Norwegian Caledonides. In 1982, he and Steven Boyer, his former PhD student, published their review of the structure, evolution and mechanical processes associated with thrust faulting, which for many years became the essential starting point for any study of thrust belts. Elliott's tragic early death in 1982, while travelling

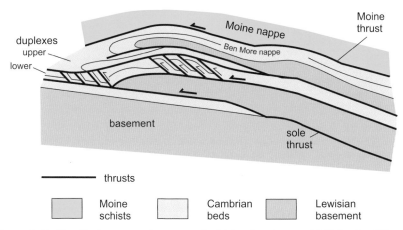

Figure 8.11 Simplified cross-section across the Moine thrust zone of NW Scotland illustrating how the upper thrust sheets have been deformed by younger movements on the lower thrusts. The term 'nappe' is traditionally used in this area to describe a thrust sheet. Modified from Elliott and Johnson, 1980.

to Zurich for a conference, cut short what would surely have been a distinguished career.

The London conference

The explosion of interest in thrust belts resulting from the North American work culminated in Britain with a conference held at Imperial College, London in 1979, from which emerged a Special Publication of the Geological Society of London in 1981. This volume included papers on thrust belts from many different orogenic belts, including several from the Canadian Rockies and one from the Appalachians; Bally himself provided an overview. The Swiss Alps were discussed by John Ramsay, the Jura by Hans Laubscher, and the Moine thrust zone by McClay and Coward.

Boyer and Elliott: the rules for thrust systems

The purpose of this key paper by Boyer and Elliott was to set out the rules governing the three-dimensional geometry of a thrust system, and to explain how, by understanding these, the time sequence of the various faults could be established. In order to do this, it was necessary to restore the deformed section to its undeformed state by deploying the section balancing methods established previously by Dahlstrom. The authors emphasize that much of the procedure and terminology they describe was already in common use by North American geologists working on the Rocky Mountain and Appalachian belts, but this paper states them in a usable form for the first time.

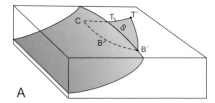 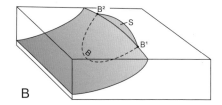

Figure 8.12 Block diagrams showing thrust terminology. **A** Diverging splay (S) has a single tip line (T) with a map termination (T´) and a branch line that intersects the main thrust surface along the line B and the erosion surface at B´; there is one corner (C). **B** Rejoining splay (S) has a branch line (B) intersecting the main thrust surface along line B and the map surface at two branch points, B¹ and B². Main thrust surface, red; splay, orange. After Boyer and Elliott, 1982.

Terminology

In order to describe the complex geometry of a jumble of connected faults, it was necessary to establish a nomenclature for the various parts of the system. These are illustrated in Figure 8.12.

- A 'thrust sheet' is a volume of rock bounded below by a thrust; it may comprise a distinctive stratigraphy that can be followed laterally for long distances.
- A 'tip line' is the margin of a thrust surface.
- A 'blind thrust' is one whose tip line does not reach the erosion surface.
- A 'cut-off line' is the intersection of a thrust surface with a particular stratigraphic horizon, and may be either a leading or trailing cut-off with respect to that horizon.
- A 'splay' is a minor thrust that meets the main thrust surface.
- A 'branch line' is where a splay meets the main thrust surface.
- Splays can be isolated, or join along a common tip line ('diverging splays'). Major thrusts often become a network of diverging splays on approaching their termination. In a 'rejoining splay', the splay meets the main thrust twice. A 'connecting splay' joins two adjoining thrusts, along two branch lines.

Thrust systems

An 'imbricate thrust' system is a family of overlapping thrust sheets that all dip in the same direction; the term is the equivalent of 'schuppen structure' used by many Alpine geologists. The individual thrusts branch upwards from a common basal sole thrust; where they diverge upwards, they are known as an 'imbricate fan' (Fig. 8.13A). Such a system is an efficient way of shortening and thickening a sequence. The individual branches of the imbricate system may end upwards as blind thrusts, or they may terminate against an upper common roof thrust. In the latter case, a structure is formed consisting of an

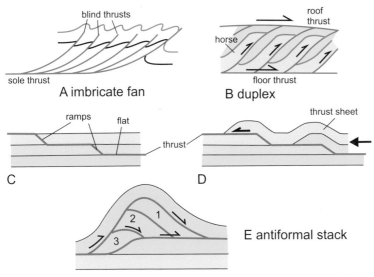

Figure 8.13 Thrust terminology. **A** Imbricate fan: note that the thrusts are replaced by folds upwards (i.e. 'blind thrusts'). **B** Duplex, consisting of floor and roof thrusts enclosing an imbricate stack consisting of several horses. **C, D** Formation of folds in the hanging wall of a thrust sheet: **C**, before thrusting, **D**, after. Note that the hanging-wall ramps become flats after thrusting. **E** Antiformal stack: successive horses (1–3) have been pushed over the flat, bending the roof thrust into an anticline. After Boyer and Elliott, 1982.

imbricate stack bounded below by a sole thrust or floor thrust and above by a roof thrust, and is known as a 'duplex' (Fig. 8.13B).

A duplex thus consists of several (often very many) individual thrust-bound units termed 'horses'. A horse is formed where a pair of thrusts diverge after branching and meet again at a further branch line. This may occur along-strike as well as in the transport direction, in which case a body of rock is completely enclosed by the thrust surfaces. Horses can be cut either from the hanging wall or the footwall of the major thrust and can consist entirely of inverted stratigraphy. The beds within a horse frequently trace out an anti-cline–syncline fold pair – hence the name! This shape is due to the way that the thrust sheet advances by climbing up an inclined thrust surface, known as a 'ramp', cut through the bed in front of it. The sheet may then follow a higher décollement surface, in which case it becomes a 'flat'. (Fig. 8.13C, D). More complex structures may be formed if successive horses climb over a flat to create an antiformal bulge in the roof thrust (Fig. 8.13E). If the duplex con-tains a simple stratigraphic sequence, it is relatively easy to calculate the total shortening involved.

Horses and duplexes are more common in the deeper, older, and more internal parts of thrust belts where thrust activity has been active for longer, whereas isolated or diverging splays and blind thrusts are more characteristic of the shallower, younger and more external parts of the system. The authors give good examples of these structures from the Moine thrust zone in the Scottish Caledonides (see Fig. 8.11), the southern Appalachians and the Rockies.

Sequence of development

As a general rule, each new thrust in a duplex system develops as a crack, which propagates from the base of a ramp and, in order to accommodate further shortening, becomes a new ramp situated forward of the last one (Fig. 8.14). The new thrust may then find a new décollement horizon at a higher level. In the case of a duplex, each new thrust sheet will carry the higher thrust sheet above itself, deforming it in the process. The fact that a particular thrust sheet is deformed into an anticline or syncline is an indication that it is older than the one below it, and confirms the forward propagation model. This principle was applied by Elliott and Johnson to the Moine Thrust in their 1980 paper, challenging the then common belief that this famous thrust was younger than the thrust sheets beneath it. Folds formed above a flat sole thrust will always be the result of the thrusting process.

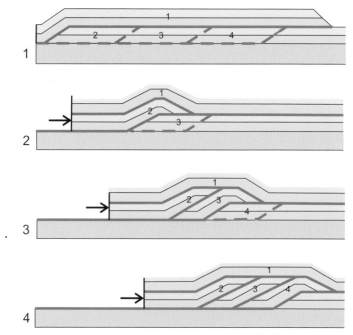

Figure 8.14 Development of a duplex. 1) Upper thrust sheet is emplaced; the positions of future thrusts are shown with dashed lines. 2) A new thrust sheet (2) forms by moving up a ramp, causing thrust sheet 1 to be deformed into an anticline. 3) Thrust sheet (3) forms in the same way as 2, by sliding up a new ramp and extending the anticlinal structure of the upper thrust sheet. 4) Thrust sheet (4) forms in the same way as 2 and 3; sheets 3 and 4 are now enclosed by thrusts – i.e. are 'horses'. Based on Boyer and Elliott, 1982.

Culminations and windows

A 'tectonic window' is an effect caused by erosion that has cut through a thrust sheet to expose lower structures beneath it. Such a structure is frequently caused by an antiformal bulge in the lower structures; this may be the result of a local thickening of the tectonic sequence, and is termed a culmination. The classic Assynt window in the Moine Thrust zone exposes a wide area of thrust sheets beneath the Moine Thrust, known as the Assynt 'culmination', which is believed to have resulted from the presence of igneous intrusions and their metamorphic aureoles within the thrust sequence.

Knowledge of how the complex geometries of thrust belts have developed can explain otherwise very puzzling outcrop shapes.

Root zones

These are the, usually steep, zones where the main thrusts of a thrust belt finally descend from view. Such zones are typically highly deformed and mylonitized, and in many cases, are located at a former subduction zone. If both margins consist of continental lithosphere, the root zone will correspond to a suture between two formerly separate tectonic plates.

Résumé

Boyer and Elliott's paper, along with a shorter account of thrust terminology by Robert Butler, which appeared in the same year, provided the essential geometric tools to enable the geometry of thrust belts to be better understood, and has been widely quoted by structural geologists since.

Subsequent developments

It was realized by structural geologists that zones of extensional structures could be analysed in the same way as compressional fold belts. This led to an improvement in the understanding of the kinematic behaviour of extensional faults and their role in the thinning of continental crust, with its implications for the development of sedimentary basins. A Special Publication of the Geological Society of London published in 1987 contains a large number of papers illustrating the wide range of studies on the subject of extensional tectonics.

Strike-slip tectonic regimes were also being subjected to the same kind of kinematic analysis. Many of the major transform faults such as the San Andreas consist of complex branching systems analogous to thrust belts, but seen vertically, and where many of the component structures were given the same names, e.g. 'strike-slip duplexes'.

Postscript

The work of Bally and Dahlstrom in the Canadian Rocky Mountains was the driving force behind a fundamental change in the way that the majority of structural geologists worldwide thought about faults and faulting – from what was an essentially dynamic process involving a brittle response to stress, to a kinematic process involving an intricate but understandable sequence of linked movements. Only by appreciating this could the complexities of the great mountain belts be unravelled.

9

Back-arc basins and trench roll-back

Historical background

The introduction of the plate-tectonic theory in the late 1960s stimulated an explosion of research activity focused on applying the new ideas to a wide range of different geological phenomena. Among the first to receive attention was the subduction process. In the simple conveyor-belt model introduced by Hess, the spreading ridges were envisaged as the sites of crustal extension, and the subduction zones of compression. However, it soon became clear from the greatly increased amount of seismic and oceanographic data that this model was greatly oversimplified. A series of oceanographic expeditions to the western Pacific undertaken by Scripps Institution of Oceanography at San Diego culminated in a paper by Dan Karig in 1970 in which he introduced the concept of the 'back-arc basin' – an extensional basin formed on the upper plate of the subduction zone. The following year, W.M. Elsasser, in discussing thermal convection, showed that the gravitational effect of a sinking cool dense slab would generate a tensional stress on the lithosphere on either side of the subduction zone. A more sophisticated model of the subduction process was clearly required.

The improved model incorporated three significant strands of evidence: 1) oceanographic data from the oceanic regions 'behind' the island arcs (i.e. between the arcs and the nearest continental margins), which demonstrated that these were, in many cases, quite young ocean basins; 2) more detailed seismic models of the subduction zone itself; and 3) a more realistic model of the stress regime across the subduction zone.

The contribution of Dan Karig

The key paper that introduced the concept of the back-arc basin to the geological community was published by Dan Karig in 1970, entitled *Ridges and basins of the Tonga-Kermadec island arc system,* in which he detailed the results of the extensive research carried out by Scripps Institution; this was followed in 1971 by a companion study of the Mariana island arc system. The most important conclusion from these two studies was that the basins in question were both young in age and extensional in origin.

The Tonga–Kermadec system

Karig's paper was based on results obtained during the 'Nova' expedition carried out by the Scripps Institution of Oceanography, employing bathymetric and seismic reflection surveys, dredged bottom sampling, drill core sampling carried out as part of the Deep-Sea Drilling Program (DSDP), magnetic profiling and heat-flow measurements.

The Tonga–Kermadec arc system (Fig. 9.1) extends from the northeast coast of the North Island of New Zealand in a north-north-easterly direction. It consists of three components, from east to west: a frontal arc, which contains the active volcanic arc; an inter-arc basin; and a third, non-volcanic, arc named the Lau–Colville ridge (Fig. 9.2).

The frontal arc

This compound feature includes the coralline islands of the Tonga archipelago, the submerged Kermadec ridge, and the active volcanic chain, which is situated on the western side of the frontal arc, and includes the volcanic islands of the Tonga and Kermadec groups. The exposed parts of the arc consist mostly of andesitic volcanic products ranging from Eocene to Pliocene in age, similar to those emitted from the presently active chain. The ridge is regarded as an uplifted block tilted to the west, exposing the oldest rocks on its eastern side. The eastern margin is marked by a steep scarp interpreted as an eastward-dipping normal fault zone (Fig. 9.3). From here, a smooth slope flattens out briefly before dropping down steeply to the trench. The upper part of the slope is covered by a thick sediment apron.

The inter-arc basin

This consists of the Lau basin in the north and the Havre Trough in the south, with an average depth of 2,500 m, separated by a submerged transverse ridge. The eastern margin of the basin is defined by steep west-dipping boundary faults, and the basin floor is characterized by a rough topography with little sediment cover. A series of linear ridges with a relief of 500–1500 m that cross the floor of the basin are interpreted as tilted fault blocks. The ridges are oblique to the margins of the basin, displaying an en-echelon arrangement, suggesting that extension has operated perpendicularly to the ridges but obliquely to the basin margins. The basin is underlain by oceanic-type crust about 7 km in thickness, contrasting with the more complex 15 km-thick crust beneath the frontal and third arcs. A weak magnetic field in the Havre trough shows NE-trending anomalies.

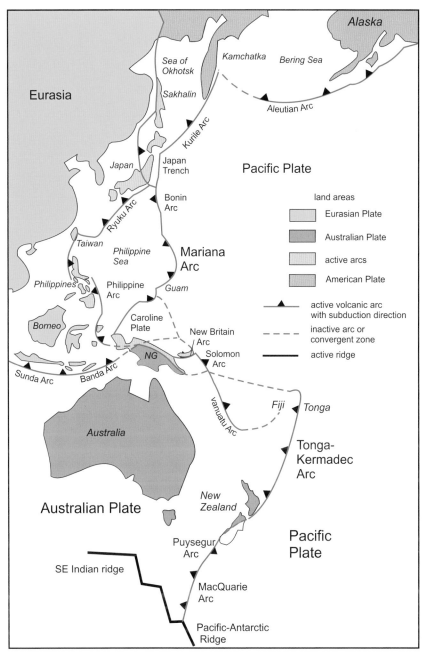

Figure 9.1 Active arc systems of the Western Pacific. Note the positions of the Tonga–Kermadec and Mariana arcs. After Leat and Larter, 2003.

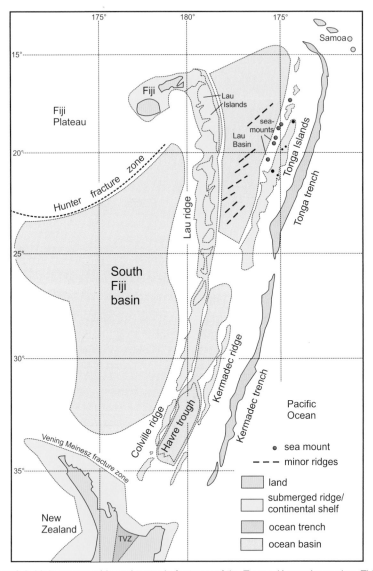

Figure 9.2 Main topographic and tectonic features of the Tonga–Kermadec region. TVZ, Taupo volcanic zone. Based on the bathymetric map of Karig, 1970.

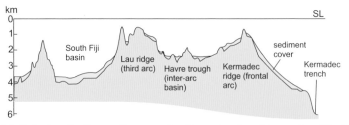

Figure 9.3 Seismic reflection profile across the Kermadec–Lau ridge sector of the Tonga–Kermadec island arc system. Vertical scale exaggerated. After Karig, 1970.

The 'third arc'

This consists of a broad submerged plateau with an average depth of 1000 m named the Lau-Colville ridge. It is an asymmetric structure with a steep eastern side interpreted as a normal fault zone similar to the eastern margin of the inter-arc basin but dipping eastwards. The western margin is a smooth slope covered by an extensive sediment apron that continues westwards into the South Fiji basin. The oldest known rocks are lower Miocene in age.

Karig's conclusions

Four main strands of evidence support the conclusion that the inter-arc basin originated by extensional rifting: 1) both ridges exhibit markedly asymmetric cross-sections, in both sediment distribution and morphology, so that both the outer flanks and inner flanks of the ridges match each other; 2) the thick sedimentary aprons that mantle the outer slopes of the ridges contrast with the extreme thinness or lack of sediments within the basin; 3) the prevalence of (apparently) normal faulting on the inner flanks of the ridges; and 4) the correlation of the structures of the island-arc system with those of the northern part of the North Island of New Zealand (e.g. the Taupo volcanic zone) which are recognizably extensional in origin.

The present chain of active volcanoes was formed no later than 5 Ma ago, and is attributed to a phase of late Cenozoic tectonism that has uplifted the eastern edge of the frontal arc and deepened the Tonga–Kermadec trench. However, the geology of the exposed islands suggests that the Tonga–Kermadec ridge has been a frontal arc since the early Cenozoic, and the Lau-Colville ridge is probably at least as old. In contrast, the thin cover of rapidly deposited sediments and rough morphology of the inter-arc basin, with relief of c.1000 m, suggests that the basin is no older than late Pliocene. Furthermore, the basin is underlain by crust of oceanic character and exhibits higher than normal heat flow. Karig concludes:

> 'An origin of the inter-arc basin by extensional rifting within an older frontal arc best satisfies the available data and suggests that the extension is related to intensification of island arc tectonism at the close of the Tertiary...' [i.e. Cenozoic]

And further:

> 'The result of such activity would be the migration of the trench-frontal arc complexes away from the Asian continent, with creation of new basins with oceanic crust on the convex [concave? (author)] sides of the frontal arcs.'

This was the critical evidence that established the nature and extensional origin of back-arc basins and was to be reinforced by the example of the Mariana arc system.

The Mariana island arc system

The Mariana arc was the second island arc system to be studied by Karig, who provides a more detailed and thorough analysis of it compared with the Tonga–Kermadec study. It proved to be an even more convincing example of his back-arc spreading mechanism. This island arc is separated from the Asian continent by the wide expanse of the Philippine Sea, which is itself bounded in the west by two further arc systems: the Philippine and Ryukyu Arcs (see Fig. 9.1). Its northern extension is named the Bonin (or more correctly, the Bonin–Izu–Ogasawara Arc). Seven separate tectonic elements are described in the Mariana system (Fig. 9.4): from east to west: the Mariana Trench; the frontal arc (the Mariana ridge); the active volcanic arc; an inter-arc basin called the Mariana Trough; a third, inactive, arc termed the West Mariana Ridge; a second, wider, inter-arc basin, named the Parece Vela Basin, which is bounded on its western side by the Palau–Kyushu Ridge. West of this last ridge is the wide Philippine Sea. The detailed topography is shown in profile in Figure 9.5.

The Mariana work was a continuation of the oceanographic program co-ordinated by the Scripps Institution using similar methods to the Tonga–Kermadec survey, together with traditional observations from the few on-land sites.

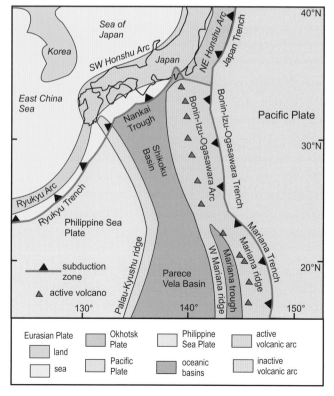

Figure 9.4 Main tectonic features of the Mariana Arc system and the adjoining island arcs. Active volcanoes shown only for the Mariana–Bonin arc system. After Taira *et al.*, 2016.

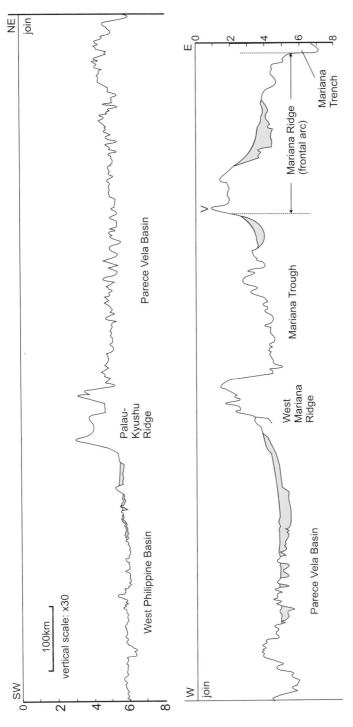

Figure 9.5 Sea-floor topography of the Mariana Island Arc system. Based on the bathymetric and seismic reflection survey of Karig (1971). V, volcanic arc. Note: x30 vertical exaggeration.

The Mariana Trench is one of the deepest parts of the global trench network with a maximum depth of over 11 km. Its western side slopes up towards the frontal arc. This is the site of the presently active subduction zone.

The frontal arc consists of the mostly submerged Mariana Ridge, from which project several islands, including Guam. The islands consist of andesitic volcanics and carbonate sediments. A seismic profile indicates fault scarps on the western flank of the frontal arc and up to 1 km of sediments resting on the volcanic basement. The mostly volcanogenic sedimentary fill dates from late Eocene to Quaternary, but is predominantly Miocene in age, and the arc appears to have been a shallow-marine ridge since the early Cenozoic, receiving volcanic detritus from active arcs to the west. Tectonic activity here was restricted to block faulting on normal faults with both N–S and E–W trends.

The active volcanic arc consists of a chain of active volcanoes situated 25–30 km west of the crest of the submerged Mariana Ridge. The sides of the ridge are defined by steep scarps formed by normal fault zones.

The Mariana Trough (not to be confused with the Mariana Trench) is a wide, roughly crescent-shaped basin about 200 km wide at its widest part, with an average depth of 4000 m. Thus the basin is 2000 m shallower than the main Pacific basin to the east, and also shallower, by 500–1,000 m, than the Parece Vela Basin to the west. The basin floor contains a series of ridges and troughs with a roughly NNE trend, arranged en-echelon, oblique to the north-trending flanking ridges. Recent pillow basalts have been dredged from some of the ridge flanks. No volcanoes have been found within the basin, suggesting that the volcanicity has erupted from fissures. The basin floor has a cover of muddy sediment containing volcanic debris up to several hundred metres thick in the troughs, but thinner on the ridges. Cores and dredges have yielded only Quaternary microfossils indicating a late Pliocene origin for the basin. Like the active arc, the Mariana Trough is also seismically active, indicating that tectonic activity is ongoing.

The West Mariana Ridge forms the western boundary of the Mariana Trough. It is a wholly submarine ridge, 50–75 km wide, extending the whole length of the Mariana Arc system and extending northwards to form the western side of the Bonin Arc (see Fig. 9.4). It is more weakly curved than the main arc owing to the crescent shape of the Mariana Trough. The crest of the ridge is around 2 km below sea level, and both margins are defined by prominent scarps. However, the ridge profile is strongly asymmetric: the eastern margin is defined by a steep scarp descending about 3 km to the basin floor, whereas on its western side, the ridge is flanked by a wide sedimentary apron. A line of seamounts is located on the western side of the ridge, and these are assumed to be the volcanic source of andesitic and dacitic material within the sedimentary sequence, which is dated as late Pliocene in age.

Reef carbonates on the ridge are of probable late Miocene to early Pliocene in age, and suggest that the arc has subsided more than 1 km since the late Miocene. The western scarp, including the seamounts, is partly covered by the uppermost sediments.

The Parece Vela Basin is similar to the Mariana Trough but much larger and deeper, and is floored by crust of near-normal oceanic thickness and character. The western part contains a series of linear basaltic ridges separated by troughs, but this ridge–trough topography becomes more subdued eastwards, giving way to a smooth broad plain formed by the sedimentary apron referred to above, which is over 1500 m thick. The sedimentary sequence consists of fossiliferous limestone of probable Oligocene to early Miocene age, succeeded by early- to mid-Miocene volcanic ash, overlain in turn by pelagic clay. Evidence of tectonic activity is limited to minor folding and faulting of the sedimentary apron over scarps bordering the basement ridges. The amount of displacement on the faults decreases upwards and the uppermost layer is draped over the scarps, indicating that tectonic activity decreased over time and finally ended before the youngest sediments were deposited.

The Palau–Kyushu Ridge forms the western boundary of the Parece Vela Basin, separating it from the wide expanse of the Philippine Sea. Like the West Mariana Ridge, this is a submarine ridge projecting up to 3 km from the basin floor, and in the section crossed by the traverse line, contains a central trough. No information about the geological composition of the ridge is recorded.

The origin of the Mariana Arc system

Several lines of evidence support Karig's conclusion that the Mariana Trough (i.e. the inter-arc basin) is extensional in origin: 1) the steep normal faults that form the boundaries of the basin; 2) the thick sediment cover on the outer flanks of the boundary ridges contrasting with the sediment-poor basin interior; and 3) the fault-block morphology within the basin, which can be traced on-shore to a horst–graben system.

The similarity in morphology, structure and sediment distribution between the western flanks of the frontal arc and the West Mariana ridge suggest a similar origin, prompting Karig to interpret the West Mariana Ridge as part of a Miocene frontal arc. Reflection profiles across the Palau–Kyushu Ridge indicate a similar but older origin.

In view of the progressive westwards thickening of the pelagic sediments in the basins of the Philippine Sea, coupled with the similar westwards increase in age of the basement, Karig suggests that these basins become progressively older westwards and that the Parece Vela Basin developed prior to the Miocene. He considers the possibility that the apparent eastwards younging of the frontal arc systems might result from an eastward 'jumping' of the trench position, trapping oceanic crust between successive arc positions. However, the fact that the volcanism represented on the Mariana Ridge commenced in the Eocene and continued through the Miocene to the present indicates that the present frontal arc has been in existence throughout this entire period, invalidating the 'eastward jump' hypothesis.

The most likely origin for the basins of the Mariana arc system was therefore some mechanism involving collapse and/or extension. Karig rules out collapse of a continental or semi-continental crust on the grounds that the similar characteristics of the whole Philippine Sea region implied that all were created by a common process, and neither dredge sampling nor seismic reflection profiling have given any indication of continental material or indeed of any pre-Cenozoic rocks. The basins appeared to be floored by 'normal', relatively young, oceanic-type crust.

A critical observation is that the Miocene apron along the western side of the West Mariana Ridge is now up to 250 km from the matching apron on the eastern side of the frontal arc. Since transport of volcaniclastic material from the position of the present active arc across the inter-arc basin seemed inherently unlikely, the source of the West Mariana apron deposits must originally have been much closer, suggesting that the original (Miocene) arc had been split apart by the formation of the inter-arc basin. This conclusion is supported by the asymmetry displayed by the ridges on either side of the inter-arc basin: the western apron on the West Mariana Ridge matches the eastern side of the main Mariana Ridge, whereas the inner flanks of these ridges are marked by steep scarps.

Karig visualizes the process beginning with a zone of crustal extension within a frontal arc, which widens until a large inter-arc basin develops. Oceanic crust then formed along the axial high of this basin, fed probably by dyke swarms. The basalt samples dredged from the axial high support this interpretation. Extensional tectonism within the inter-arc basin is indicated by the steep normal faults on the flanks of the ridges and is also supported by the occurrence of shallow-focus earthquake activity. The high shear-wave attenuation (see later) suggests that the basin is underlain by anomalously warm mantle.

The absence of volcanogenic material in the Miocene limestones of the frontal arc, and the lack of sediments older than late Pliocene within the Mariana Trough, indicate that the presently active arc system commenced in the Pliocene with the formation of a trench, followed by a volcanic arc, accompanied by uplift of the frontal arc and minor compressional effects. These would have been followed in turn by the gradual expansion of the inter-arc basin. Prior to this phase (i.e. in the Miocene) the frontal arc was probably straighter, and aligned with the Bonin system to the north; the expansion of the inter-arc basin would have been responsible for its present curved shape.

The opening of the Parece Vela Basin to the west is assigned to an earlier phase of early Miocene (or perhaps late Oligocene) extension, with the central zone of rough topography marking the position of a former axial high. A still earlier phase may have been responsible for the West Philippine Basin.

There is evidence for substantial vertical movements in both the marginal basins and the ridges. From the axial high of the Mariana Trough, at approximately 3 km depth, the basins increase in depth westwards to 6 km in the West Philippine Sea, and this depth increase is accompanied by an increase in age, suggesting that the oceanic crust of the basins has subsided with time, after its formation along the axial highs. The West Mariana Ridge has subsided by more than 1 km and the Palau–Kyushu Ridge (assuming it was originally at sea level) by approximately 3 km.

Karig concludes that his interpretation is consistent with the sea-floor spreading model of ocean crust being created at shallow depths along the axis of a basin that subsides with time as it cools, implying that the aseismic ridges, or remnant arcs, subside passively along with the basins. He states finally that his interpretation:

> 'does no irreparable damage to recent hypotheses which assume a plate-like nature of the Earth's crust and upper mantle…but certainly requires modification of assumptions concerning simple plate boundaries…'

The next stage in the evolution of ideas on the origin of back-arc basins required further investigation of the subduction mechanism.

Slab roll-back

The first suggestion of a possible mechanism that could explain the occurrence of extension on the upper plate of a subduction zone came from a 1971 paper by Walter Elsasser. Elsasser was primarily concerned with how the sea-floor spreading mechanism would affect thermal convection in the deeper Earth. He starts with the presumption that the relative weakness of the asthenosphere compared with the lithosphere, owing to its much lower viscosity, means that no strong horizontal stress can be transmitted from the asthenosphere to the lithosphere. It follows that strong horizontal stresses observed in the lithosphere must have been generated there and do not result from forces within the asthenosphere, such as shear forces generated by convective movements. This places a constraint on how a mantle convection system can operate and suggests that it must be the movements of the lithosphere plates that control the upper parts of the convective system.

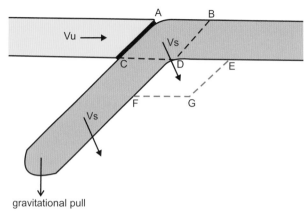

Figure 9.6 The roll-back mechanism. Elsasser's explanation of slab roll-back: the effect of gravity on an oceanic lithosphere slab that is denser than the surrounding asthenosphere is to exert a downward force on all parts of the subducting slab, such that the velocity of the inclined part of the slab (Vs) tends to move the slab downwards and backwards, so that, e.g., the section ABCD moves to a new position DEFG. If the velocity of the upper slab (Vu) is less than the horizontal component of Vs, the upper slab will experience extension. After Elsasser, 1971.

Elsasser goes on to note that much of the lithosphere appears to be under tensional rather than compressional stress, and argues that the gravitational force exerted by the sinking slab at a subduction zone must be the main influence on the stress state of the oceanic plates. This force has come to be known as 'slab pull' although Elsasser himself does not use this term. In a diagram reproduced in Figure 9.6, he shows that this slab-pull force may also exert an extensional force on the upper plate, by causing ocean-ward retreat from the initial position of the trench, thus providing a mechanism for Karig's back-arc spreading. This process of slab, or trench, retreat is now known as 'slab roll-back'. Although subsequent work has shown that the mechanism of subduction is much more complex than is envisaged in Elsasser's simple model, there is no doubt that the roll-back idea has been extremely influential.

Later refinements to the subduction model
The evidence from seismicity

The Lamont-Doherty geophysicists continued their work on global seismicity that had been so influential in establishing the plate-tectonic model in 1968 (see chapter 6), and by 1973 a greatly improved picture of the seismicity pattern around subduction zones had emerged.

Down-dip stresses

Useful information about the subduction process can be obtained by studying the 'focal mechanisms' of earthquake foci located within the downgoing slab.

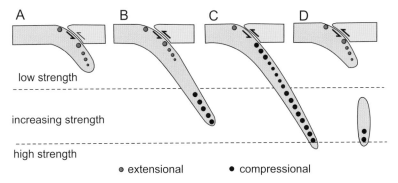

Figure 9.7 Diagram explaining the distribution of down-dip stress within subducting slabs. In **A**, the slab sinks into the asthenosphere under the load of its excess mass, and is wholly in extension. In **B**, the subducting slab has penetrated further and has met with increasing resistance due to the increasing strength of the mantle; here the lower part of the slab is in compression. In **C**, the slab has penetrated further into the higher-strength mesosphere, and the whole slab is under compression. In **D**, a piece of slab has broken off; the upper part has reverted to the extensional state of case A, while the lower part is still under compression. The size of the circles indicates the relative amount of seismic activity. After Oliver *et al.*, 1973.

By analysing the 'first motion' at the source of an earthquake it is possible to decide whether the initial down-dip stress at that point is either compressional or extensional. When Bryan Isacks and Peter Molnar applied this technique in a comprehensive worldwide study of 24 separate subduction zones in 1971, some interesting results were revealed, summarized in Figure 9.7. Some slabs exhibit only down-dip compressive stress solutions, indicating that the slabs are in a state of compressive stress throughout, while others appear to be wholly in extension. A third group give mixed solutions, indicating extension in the upper part and compression in the lower, with a gap in the middle.

This study confirmed the view that the subduction process was controlled by the gravitational pull on the denser, cooler subducting slab, but only until the slab reached a depth where the increasing resistance of the higher-strength mantle material towards the base of the asthenosphere produced a state of down-dip compression within, first, the lower part of the slab, and later, throughout its length. These results appeared to confirm Elsasser's views of the importance of slab-pull, at least in the early stages of the subduction process.

Seismic wave properties

Seismic wave propagation, especially for shear waves, is more 'efficient' within the slab and the seismic velocities are higher than is the case within the surrounding mantle. The efficiency, or effectiveness, of shear waves is measured by the 'attenuation' of the wave train – i.e. how long it takes for it to die out. This property is denoted by the symbol 'Q': high-Q means a low

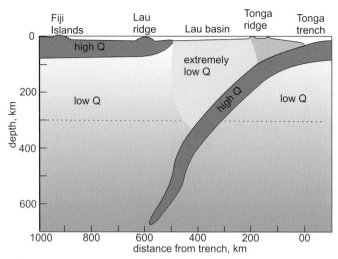

Figure 9.8 Schematic section across the Tonga arc showing the lithosphere plates, characterized by high Q, contrasting with the low-Q asthenosphere and the ultra-low-Q zone beneath the Lau basin. After Oliver *et al.*, 1973.

degree of attenuation (a long wave train) and low-Q means a high degree of attenuation, or short wave train. In 1971, Muawia Barazangi and Bryan Isacks applied this analysis to the Tonga Arc system, and found that the asthenosphere can be described in terms of a low-Q layer extending normally to a depth of 300–350 km; however, in a subduction zone, this low-Q zone rises much closer to the surface and can be distinguished from the slab, which is characterized by both higher Q and higher seismic velocities (Fig. 9.8). This is consistent with the Karig model requiring the existence of warmer asthenospheric mantle material able to generate basaltic melts beneath an extensional back-arc region.

Dynamics of the downgoing slab

The Isacks and Molnar study on down-dip stress in subducting slabs was followed in 1973 by an influential paper by a team (now based at Cornell University) consisting of Jack Oliver, Bryan Isacks, Muawia Barazangi and Walter Mironovas entitled *Dynamics of the downgoing lithosphere*, in which they discuss the conclusions that can be drawn from the analysis of focal mechanisms and seismic-wave attenuation.

The authors consider various possible types of slab descent geometry based on the distribution and focal mechanism of earthquakes around subduction zones. These are summarized in Figure 9.9, in which various possibilities are considered, including slabs sticking at the boundary of the lower mantle, and bits of a slab breaking off to leave a trail along the boundary. Their conclusions are as follows:

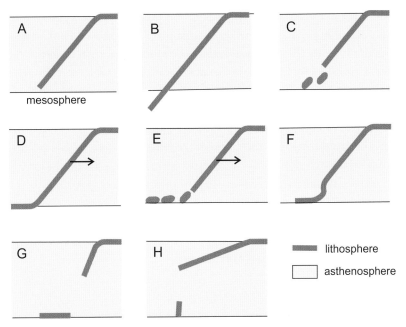

mesosphere

lithosphere

asthenosphere

Figure 9.9 Schematic sections illustrating possible configurations of subducting slabs. **A** The slab has penetrated to the base of the asthenosphere, where it has been assimilated into the mesosphere. **B** The slab has penetrated through the asthenosphere and the topmost part of the mesosphere. **C** Sections of the lower end of the slab have broken off. **D** Roll-back has caused the slab to migrate to the right, and the lower end has been unable to penetrate into the mesosphere and lies along the base of the asthenosphere. **E** As **D**, but the lower part has broken into sections. **F** The lower part of the slab has become contorted due to its inability to penetrate the mesosphere. **G** and **H** The lower part of the slab has sunk rapidly, become detached, and lies either horizontally or vertically along the base of the asthenosphere. These are based on real examples of deep seismic source configurations. After Oliver *et al.*, 1973.

1 Downgoing slabs are generally continuous.

2 Downgoing slabs are more dense than the surrounding mantle and sink under their own weight.

3 Downgoing slabs penetrate the upper mantle easily for the first few hundred kilometres but encounter increasing resistance below that level.

4 There is no conclusive evidence of slabs penetrating below *c.*700 km, and some inconclusive evidence that they are unable to do so.

5 In a few cases, there is evidence of lower portions of a slab having become detached from the upper part and thus leaving a gap – perhaps because at one time they had sunk at a greater rate than the upper part; this suggests that laterally migrating arcs may leave a trail of detached slabs whose contrast with the surrounding mantle decreases with the age of detachment.

6 The present (inconclusive) evidence suggests that 700 km is the lower boundary of relatively rapid mantle flow.

The work of Oliver and his colleagues on the stress state in sinking slabs was important in persuading geologists that plate motion could not be simply explained either by the pulling action of the slab or the push from the ridge, but was controlled by the interaction of several primary forces. Although the orientation of the maximum horizontal stress axis across a subduction zone is generally parallel to the convergence direction of the opposing plates, this stress state is often confined to the volcanic arc itself, and is replaced by an extensional stress field in the back-arc region. However, where no active opening or spreading of a back-arc exists, as in the case of the Peru–Chile, Japan and Kurile subduction zones, the compressive stress appears to be transmitted directly to the interior of the over-riding plate.

In order to understand how a state of extensional stress can exist in the tectonic environment of a subduction zone, it was necessary to examine all the possible forces that could impact on the zone.

Plate-boundary forces

The various types of force acting at plate boundaries were summarized by Forsyth and Uyeda in 1975 and amplified by Bott and Kusznir in 1984. The more important of these are as follows:

- **Slab-pull** This force acts on a subducting oceanic plate and results from the negative buoyancy of the cooler, denser lithosphere of the descending slab. This is potentially the largest of the forces acting at a subduction zone but is partially counteracted by resistance forces produced by the down-bending and collision processes.

- **Subduction suction** This force, originally termed 'trench suction' by Elsasser (1971), is caused by the pulling action of the subducting slab on the upper plate of the subduction zone. Both slab-pull and subduction suction can produce a component of extensional stress acting on the adjacent lithosphere, provided that the resistance forces are sufficiently low. The size of these resistance forces is highly dependent on the length (i.e. age) and velocity of the subducting plate.

- **Ridge push** This force acts at an oceanic spreading ridge to push two plates apart. It results from the positive buoyancy created by the extra mass of hotter, less dense material underlying the ridge, and produces a lateral compressive stress on the plates on either side.

- **Mantle drag** This is a shear force acting at the base of the moving plate. Depending on the direction of any convective movement within the asthenosphere, this shear force may either resist the movement of the plate above, or enhance it. However, due to the lower viscosity of the asthenosphere compared to that of the lithosphere, the size of the mantle drag force is considered to be small in comparison to the other forces mentioned.

- **Resistance forces** All the above forces can be opposed by frictional resistance forces produced along the interface between two bodies moving in opposite directions or at different speeds.

Relevance to back-arc basins

Extensional tectonic regimes are found in both continental and oceanic crust on the upper plates of subduction zones, as was evident in the case of the Tonga–Kermadec and Mariana arcs discussed above. The question to be decided is what the critical factor, or factors, are in determining that a state of extension should exist in some subduction zones and not in others. One suggestion was that the angle of subduction was important in determining the state of stress in the upper plate: that shallow-dipping slabs were associated with compressional stress, and steeply dipping slabs with extensional stress, on the upper plate. Since the upper plates are the sites of enhanced heat flow associated with the production of volcanic arcs, it is reasonable to assume that in such environments the upper plates would be weakened and thus more susceptible to failure under a comparatively small extensional stress. Although the stress generated by the subduction process is applied equally to both plates, only the thermally weakened upper plate would fail.

Factors affecting subduction geometry

The various factors affecting the geometry of subduction zones were analysed by Cross and Pilger in 1982 (Fig. 9.10). These authors recognize four interdependent factors: 1) rate of relative plate convergence; 2) absolute velocity of the upper plate towards the trench; 3) age of the subducting lithosphere; and 4) presence or absence of obstacles such as seamounts or oceanic plateaux on the subducting plate.

Convergence rate

Cross and Pilger concluded that large convergence rates are associated with a low subduction angle and small convergence rates with a steep angle (Fig. 9.10A).

Absolute upper-plate velocity

Absolute motion of the upper plate is an important factor independent of the relative convergence rate (Fig. 9.10B). Since the trench is fixed to the upper plate, it must move with it. Thus absolute motion of the upper plate towards the subduction zone will cause the upper plate to over-ride the trench, decreasing

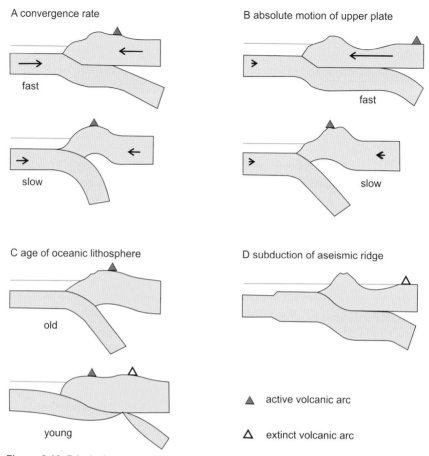

Figure 9.10 Principal controls on subduction geometry. Schematic profiles illustrating the effects of: **A** varying convergence rate; **B** varying absolute upper-plate motion; **C** varying age of subducting oceanic lithosphere; and **D** the subduction of aseismic ridges, on both the inclination of the subducting slab and the distance between the trench and the volcanic arc. After Cross & Pilger, 1982.

the subduction angle. Conversely, movement of the upper plate away from the subduction zone (or very slowly towards it) will have the effect of dragging the trench away from the sinking slab, which could steepen the slab, or ultimately rotate it backwards.

Slab age

The influence of the age of subducting oceanic lithosphere is a consequence of the increased thickness and average density, together with the increased depth of the sea floor, as the lithosphere moves away from the spreading centre. This has the effect of increasing the negative buoyancy and consequently of steepening the dip of the sinking slab, other factors being equal (Fig. 9.10C). This correlation of lithosphere age with slab dip is particularly evident when comparing the subduction zones of the western and eastern Pacific, as noted by Peter Molnar and Tanya Atwater in their 1978 paper. These authors point out that all the subducting lithosphere along the subduction zones bordering North and South America is less than 50 Ma old, whereas in the island arcs of the western Pacific, such as the Tonga, Mariana–Bonin, Japan and Kurile zones, the subducting lithosphere is greater than 100 Ma old. It is the latter zones that are associated with back-arc basins, suggesting that the ease of sinking of older, colder lithosphere is a critical factor in causing seaward migration of the trenches, leading to extensional conditions and back-arc spreading on the upper plate.

The authors note that in the zones of the Eastern Pacific, where there are currently no back-arc basins, the upper (Americas) plate is associated instead with Cordilleran tectonics, characterized by high mountain ranges and broad zones of deformation. They conclude that it is the subduction of the young, warm lithosphere that produces the typically low subduction angle in these zones, and suggest that, in order for the slab to sink, it may require a larger compressive force acting across the subduction zones, which manifests itself in the deformation of the upper plate.

Obstacles on the subducting plate

The presence of obstacles such as aseismic ridges, oceanic plateaux and seamounts on the subducting oceanic plate was another factor influencing the behaviour of the sinking slab (Fig. 9.10D). All these structures consist of regions of crust that are both topographically elevated and of reduced mean density. The result of attempting to subduct such structures is that the angle of subduction is reduced, in some cases to the point where the slab dip is so shallow that volcanism is extinguished above it. Several examples of this occur along the trenches bordering South and Central America. In the western Pacific, it is notable that in several cases the cuspate junction between adjacent arcs coincides with an aseismic ridge, suggesting that subduction may have been slowed down or even halted at these points, with the consequence that back-arc spreading has been confined to the sectors between the ridge intersections.

Later developments

By the mid-1980s, the main elements of a workable theory of subduction were in place. It was clear what conditions were required to enable an extensional environment to develop on the upper plates of subduction zones, with the consequent emergence of back-arc basins. Further studies of the numerous other present-day examples of back-arc basins had shown that Karig's model was applicable worldwide, although studies of the magnetic stripe pattern in several examples indicated a more complex pattern than was first envisaged (e.g. see Weissel, 1981). Attention now shifted to applying the model to oro-genic belts. One of the first to realize the implications of the back-arc spreading model to the interpretation of old orogenic belts was John Dewey who, in 1980, developed a general theory linking ophiolite occurrences to the new subduction model. It was Dewey who introduced the term 'slab roll-back' to describe the mechanism first suggested by Elsasser a decade earlier.

10

Hot-spots and mantle plumes

Historical background

The introduction of the plate-tectonic theory in 1967–68 gave a satisfactory explanation of the relationship between the formation of new crust at ocean ridges and destruction of old at trenches, and linked this to a model of mantle convective circulation. However, it failed to explain the formation of 'intraplate' volcanic centres such as Hawaii that were situated far from the nearest plate boundary. Moreover, it became clear that the location of much of the plate-boundary network was continuously changing on a geological timescale, and that there could not be a simple geometric relationship between the mantle convection system and the plate boundaries.

The method for determining plate movement vectors given by McKenzie and Parker in their 1967 paper and used by Vine and Hess in their map of worldwide plate movements (see Fig. 6.13) gives only movement vectors of the plates relative to an Antarctic plate assumed to be stationary. However, Tuzo Wilson in 1963 had suggested a method for determining 'absolute' plate motion by noting that, at a number of locations over the Earth's surface, volcanic activity appeared to have been concentrated over long periods of time. He called these areas 'hot-spots' and identified several including Hawaii and Iceland. He showed that the movement of a plate over these hot-spots would produce a linear chain of volcanic islands becoming progressively older away from the current centre of activity.

The contribution of Tuzo Wilson

Wilson's contribution to the plate-tectonic theory has already been discussed in chapter 6, where his identification in 1965 of the role of transform faults was the key discovery that led directly to the plate model. However, this discovery was foreshadowed in an earlier paper published in 1963 entitled *Hypothesis of Earth's behaviour* in which Wilson lays out much of the groundwork for the plate-tectonic theory. He starts with the recognition of a strong lithosphere, capable of sustaining significant vertical movements, resting on a weaker asthenosphere in which flows of the order of centimetres per year are envisaged. The question he then asks is whether this strong lithosphere is also capable of horizontal movements that can produce continental drift by means of rifting in some places and compression and over-riding in others, noting that, for example, some large faults such as the Great Glen in Scotland and the San Andreas in California have caused displacements of 'tens or hundreds of miles', and that 'great scarps [i.e. escarpments (author)] on the ocean floor offset magnetic anomalies by as much as 750 miles'.

The significance of the volcanic island chains

Wilson goes on to discuss the case of several chains of oceanic islands that increase in age with distance from the East Pacific Rise, including the Hawaii, Society, Tuamotu-Gambier, to the west of the ridge, and the Galapagos and Easter-Sala y Gomez to the east (Fig. 10.1). The specific case of Hawaii had been the subject of another paper by Wilson, also published in 1963, entitled *A possible origin of the Hawaiian islands* in which he noted that the ages of these islands ranged from 70 Ma at the far end of the chain, near the Aleutian island arc, to the present location of volcanic activity in the Hawaiian Archipelago.

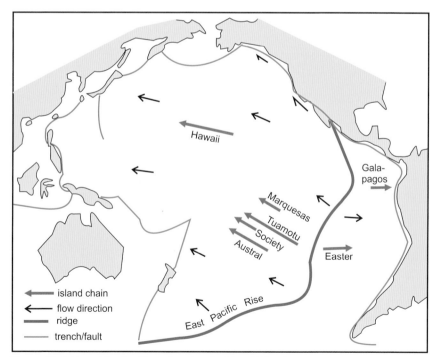

Figure 10.1 Map of the Pacific region showing Tuzo Wilson's concept of the volcanic island chains indicating the flow direction of the Pacific Ocean floor. The island chains become older in the direction of the arrow. After Wilson, 1963b.

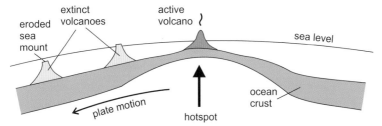

Figure 10.2 Wilson's explanation of the origin of the Hawaii island chain. Schematic diagram to illustrate the creation of a hotspot trail by a plate moving across a fixed hotspot. Not to scale. After Wilson, 1963b.

He proposed that this arrangement could have resulted from the activity of a relatively fixed mantle hot-spot over which the Pacific plate had moved over time, leaving its track in the form of a chain of volcanic islands (Fig. 10.2). The prominent bend in the chain (see Fig. 10.5) was attributed to a change in the movement vector of the Pacific plate at *c*.35 Ma.

Wilson also identified similar aseismic ridges on each side of the mid-Atlantic ridge: for example, the Tristan da Cunha–Walvis ridge directed towards Southwest Africa, and the corresponding Rio Grande ridge towards South America (Fig. 10.3). He concludes:

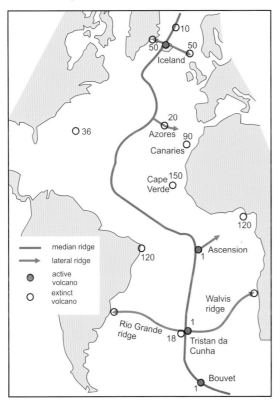

Figure 10.3 Map of the Atlantic Ocean showing the median ridge and the lateral ridges identified by Wilson. Active or recently active volcanoes are situated at or close to the ridge axis, whereas the extinct volcanoes become older the further they are from the axis. Numbers indicate age of volcanicity in Ma. The localities indicated by the red circles are all now regarded as hot-spots. After Wilson, 1963b.

'If, due to the high heat flow on the mid-ocean ridge, volcanics have long been active in about the position of Tristan da Cunha, horizontal currents would have carried volcanic piles successively off the ridge where they formed. They would have been detached from their source and (are) hence inactive... [He goes on to note that]... the ends of the Rio Grande and Walvis lateral ridges are exactly opposite points on the coasts of South America and Africa which would fit on the basis of the match of the shape of the shorelines. What might otherwise be a remarkable coincidence becomes a natural consequence if these continents had once been together with a volcanic source between them.'

It was noted by Wilson in a subsequent paper in 1973 that the present mid-Atlantic ridge is located 400–500 km west of the currently active hot-spot at Tristan da Cunha, suggesting that the plate boundary has moved west relative to the hot-spot (Fig. 10.4).

Other examples

Wilson proceeds to state that these observations give a method of fitting continents back together as an alternative proof of continental drift, and extends the argument to the North Atlantic, where similar connections exist from Iceland to Greenland in the west and Europe in the east. He suggests that continental drift could have moved Greenland away from North America, opened the Siberian basin, and compressed the Verkhoyansk Mountains in a rotational movement with a 'pivot of rotation' lying at the New Siberian Islands.

Other examples were suggested: for example, the aseismic ridges of Amsterdam–Cape Naturaliste, and Amsterdam–Kerguelen–Gaussberg record the separation of Australia from Antarctica, and the Chagos–Maldives–Laccadive ridge in the Indian Ocean track the northward movement of India.

However, the rifts that split Gondwana and result in the Atlantic, Indian and circum-Antarctic Oceans are situated centrally (or near-centrally) in their respective oceans, whereas the East Pacific Rise lies on the eastern side of the Pacific. Consequently the greater flow is to the west-northwest, as indicated by the orientation of the Hawaiian and Society island chains. He concludes that:

'The ocean floors must therefore be disappearing in trenches lying 'at right angles' to this direction, such as the East Asian arcs, the Tonga–Kermadec Islands and, in the east, the Andes [i.e. the Peru–Chile trench (author)]. These are the most active seismic regions in the world with most of the deep earthquakes.'

It is from the specific case of the aseismic ridges of the western Pacific that

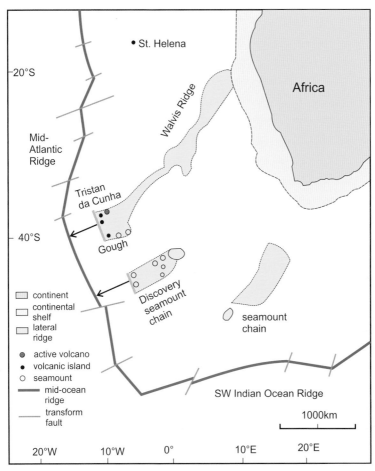

Figure 10.4 Wilson's map of the South Atlantic showing the volcanic centres of Tristan da Cunha, Gough, and Discovery seamounts offset from the mid-Atlantic Ridge, suggesting that the ridge has moved westwards away from the hotspots. After Wilson, 1973.

Wilson develops his idea of hot-spots, leading directly to the plume concept. He observes, in a key paragraph near the end of his paper, that:

> 'These chains are straight and extend in one direction only from an active or recently active volcanic centre which is not a median ridge ... [and] ... an explanation is possible if the central parts of convection cells in the mantle move more slowly below about 200 km than above and if the sources of lava under the active ends of the chains lie so deep.'

What are hot-spots?

As a result of Wilson's hypothesis, it soon became clear that a large number of possible examples of what came to be known as 'hot-spots' could be identified:

estimates vary from 40 to 150, and there is no generally agreed definitive list. The characteristics of a hot-spot are usually considered to be the following.

1 A topographic swell (relative to its surroundings) of 500–1200 m height, and 1000–1500 km width exists, which must be supported by a mass of less dense, presumably warmer, mantle material, giving it gravitational stability.

2 The topographic swells are capped by active or recently active volcanoes.

3 Most swells are associated with medium-scale gravity highs, indicating the presence of a mass excess at depth (however, some are also situated on much larger-scale gravity lows).

4 Many swells (especially in the oceans) are located at the end of an extinct volcanic chain – i.e. a hot-spot 'trace' or track.

5 Swells exhibit high heat flow, indicating the presence of abnormally warm mantle beneath.

Former hot-spots are characterized by concentrations of igneous activity, often termed 'large igneous provinces' (LIPs). However, regions of volcanic activity associated with subduction zones (i.e. subduction-related volcanic arcs) are not considered to be hot-spots.

Examples of hot-spots

Hawaii was the first and most obvious example of a hot-spot, but Iceland is another, which is equally well known and extensively documented. Many hot-spots, such as Iceland, lie on ocean ridges; others are situated on continental crust: Yellowstone Park in Wyoming is a well-studied example of the latter. There is no general agreement over the status of many of the hot-spots that have been proposed, but the map in Figure 10.5 shows the 17 well-established examples used by Morgan in his 1972 paper (see below).

The plume concept

Wilson's idea was that a hot-spot was the surface manifestation of a deep-seated body of warmer mantle material, later to be termed a 'plume', rising from depth and forming partial melts near the surface to provide the source of the volcanicity. The interesting question that then arose was how the plumes could be integrated into a theory of mantle convection. By the early 1970s the plate-tectonic theory had been widely accepted and many geologists commenced thinking about how it could be applied to various geological problems.

The contribution of Jason Morgan

Jason Morgan had made an important contribution to the plate-tectonic theory in his 1968 paper in which he described the movement of the crustal blocks

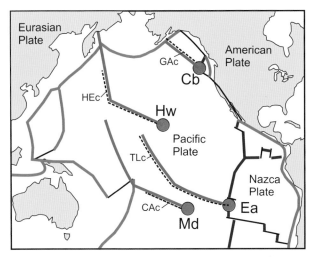

Figure 10.5 Morgan's test of the fixed hot-spot hypothesis, using four fixed hot-spots: Cobb, Hawaii, MacDonald and Easter. The four green lines were produced by rotating the Pacific Plate backwards in time over the 'fixed' hot-spots: first 34° between 0 Ma and 40 Ma, then 45° between 40 Ma and 100 Ma. The green lines follow the hot-spot trails as shown by the volcanic island chains (dashed lines). Hot-spots: Cb, Cobb; Ea, Easter; Hw, Hawaii; Md, MacDonald. Island chains: CAc, Cook-Austral; GAc, Gulf of Alaska; HEc, Hawaii-Emperor; TLc, Tuamotu-Line. Red lines, subduction zones; blue, ridges; black, faults. After Morgan, 1972.

(i.e. plates) as rotations about poles of rotation (see chapter 6 and Fig. 6.13). In three key papers published between 1970 and 1973, Morgan developed Wilson's ideas into a hypothesis of mantle behaviour in which the concept of a deep-mantle 'plume' played a central role.

In his 1972 paper, entitled *Deep mantle convection and plate motion*, Jason Morgan introduced additional lines of evidence to support the idea of plates moving over 'fixed' hot-spots. He took four examples of hot-spots: Juan da Fuca, Hawaii, Easter Island and MacDonald seamount. These are located at the ends of chains of volcanic islands and seamounts: the Gulf of Alaska chain ending at Cobb Seamount, the Hawaii–Emperor chain ending at Hawaii, the Austral–Gilbert–Marshall chain ending at Easter Island, and the Tuamotu-Line chain ending at MacDonald Seamount (Fig. 10.5). He rotated the Pacific plate backwards in time over these supposedly fixed hot-spots: first by 34° about a pole at 67°N 73°W (from 0 Ma to 40 Ma ago) then 45° about a pole at 23°N, 110°W (between 40 Ma and 100 Ma ago). The four lines generated are near parallel to each other and also to the trends of the hot-spot chains, thus providing convincing evidence to support the theory that the plates moved as internally undeformed entities, first in a NNW direction from 100 Ma to *c*.40 Ma ago, then in a more WNW direction up to the present. Moreover the hot-spots over which they moved appeared to be (relatively) stationary over timescales of the order of

100 Ma. Morgan qualified his conclusions by stating that if the hot-spots were allowed to migrate by up to about 0.5 cm per year, a better fit to the data could be achieved. This is such a small proportion of the velocity of the Pacific plate (c.7 cm/yr) that the hot-spots could still be considered stationary over these timescales.

The second argument deployed by Morgan is based on the palaeomagnetic pole positions determined for the seamounts. These show a reasonably close fit to the apparent polar-wander track for the Pacific plate over the last 100 Ma, which, assuming that the magnetic pole has not wandered far from its present position, reinforces the argument that the hot-spots did not move relative to a fixed Earth frame.

Morgan then calculates present-day plate movement vectors relative to the hot-spots. Using the relative plate movement vectors determined by spreading rates and transform faults from his 1968 paper (see chapter 6 and Fig. 6.13), Morgan added to each vector the amount of rotation required to move the Pacific plate back to the hot-spot positions. Assuming that the hot-spot frame really is fixed, this exercise should produce the 'absolute' movement vectors for each plate as shown in Figure 10.6.

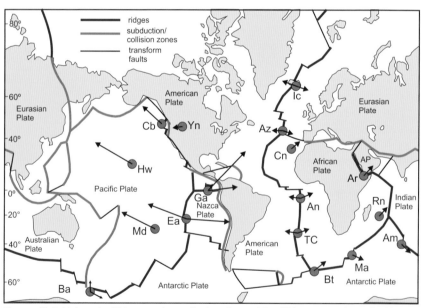

Figure 10.6 Hot-spots. This map shows the locations of the 17 hot-spots shown by Morgan in his 1972 paper. Many more have been suggested subsequently, but there is no generally agreed list. The arrows show the velocities of the plates relative to a 'fixed' hot-spot frame. Am, Amsterdam; An, Ascension; Ar, Afar; Az, Azores; Ba, Balleny; Bt, Bouvet; Cb, Cobb seamount; Cn, Canaries; Ea, Easter; Ga, Galapagos; Hw, Hawaii; Ic, Iceland; Ma, Marion; Md, MacDonald; Rn, Réunion; TC, Tristan da Cunha; Yn, Yellowstone. After Morgan, 1972.

The nature of plumes

After setting out the case for a fixed hot-spot framework, Morgan proceeds to examine the nature of the body of mantle material, for which he uses the term 'plume', that supplies the hot-spots. The term plume does not appear in Wilson's 1963 paper and it is not clear who first proposed it, but from the time of Morgan's three papers onwards, it has become the accepted way of describing these bodies. He envisages a series of relatively narrow cylindrical bodies that expand into a mushroom shape beneath the hot-spots, and suggests that they represent part of a lower-mantle convective system that provides the motive force for continental drift. About 20 of these plumes were thought to bring heat and 'primordial material' up into the asthenosphere, there to induce horizontal currents radiating away from the plume head. The return flow was believed by him to be uniformly distributed throughout the mantle.

Morgan presents two main arguments in support of this model.

1   Most hot-spots are located either near ridge crests or on continents and were active prior to continental break-up.
2   The association of hot-spots with gravity and topographic anomalies means that plumes must generate significant stresses. Estimates of the magnitude of these stresses are comparable to those associated with ridges and trenches.

Hot-spots that do not fit this simple model are of two kinds: areas of active volcanicity located at some distance from the relevant hot-spot, and oceanic volcanic centres situated far from any current plate boundary. The former are ascribed to 'delayed activity' from magma bodies originally supplied from the plume and carried laterally along with the plate. The latter, which includes Hawaii itself along with the MacDonald Seamount, the Canary Islands and also Yellowstone on the North American continent, may have been formed when the present Pacific ridge system was in a different position.

Some hot-spots were clearly active before the present ocean ridges were formed, and are thought to have been responsible for continental break-up. Examples cited are the Bouvet hot-spot, between southern South America and southern Africa, responsible for Jurassic volcanicity in Patagonia, the Tristan da Cunha hot-spot in the South Atlantic for the flood basalts in the Parana basin in Brazil, and Iceland in the North Atlantic, where the early Cenozoic volcanicity in East Greenland and NW Scotland are ascribed to the Iceland plume (see Fig. 10.6). Morgan takes this as evidence that the mid-Atlantic ridge was formed as a result of the continental break-up of Pangaea along the line of the Atlantic plumes.

The significance of the gravity and topographic anomalies

Morgan states that the isolated gravity highs situated over Iceland, Hawaii and most of the other hot-spots are 'symptomatic of rising currents in the mantle … [and although] … the less dense material in the rising plume produces a broad negative anomaly, the net gravity over the rising current is positive'. The mid-ocean ridges are exceptionally shallow near the hot-spots and Morgan points out that this topographic high is 'another manifestation of the rising plume'. (In fact, more recent maps of the global 'geoid' anomaly pattern (see Fig. 10.8) show little correlation between hot-spots and regional large-scale negative anomalies).

Plume magnitude

Morgan calculates that a positive anomaly of 20 mgals and a topographic excess of 1 km would be produced by a cylindrical mass 150 km in diameter and 1000 km in depth. He compares these figures with the stresses computed for spreading ridges and trenches: 300 bars being the upper limit for ridges and 100 bars for trenches. A plume of this size, with an upward velocity of 2 m per year, beneath a lithosphere 70 km thick, over an asthenosphere 200 km thick with an average viscosity of 3×10^{21} poise* would generate a shear stress of $c.100$ bars at a distance of 500 km from the plume. This is of the same order of magnitude as the calculated stresses for ridges and trenches, but is predicated on an estimate of plume size and shape for which there was, as yet, no evidence. [*the poise is the cgs unit of viscosity; the viscosity of water=0.01 poise.]

The ridges are not stationary

The spreading ridges on three sides of the African plate must be moving apart, pushing Africa north-eastwards, as there is no subduction taking place to consume the extra oceanic material created. Similarly, as the mid-Atlantic ridge is (relatively) fixed, since all the Atlantic hot-spots are near the present crest, the mid-Indian Ocean ridge must be migrating eastwards faster than the African plate is moving north-eastwards. Morgan argues that this ridge would have migrated over the Réunion plume, which must once have been located beneath the Indian plate, hence providing the voluminous volcanicity of the Deccan traps.

From the above evidence that ridges clearly migrate, and also from the fact that spreading is symmetrical on either side of the ridge axes, Morgan concludes that 'rises [i.e. spreading ridges (author)] do not drive the plates'. Instead, he favours a model where material from the asthenosphere rises passively into the void created as the plates move apart. Since the weakest part of the ridge is the central warmest part, where the previous batch of magma has arisen, each new dyke will be inserted there, imparting a symmetrical spreading pattern.

Since the asthenospheric material drawn into the ridge axis would come from relatively shallow depths, there need be no connection with deeper-level conditions at the base of the asthenosphere.

Ridge-axis offsets along transform faults are another factor that is difficult to explain if spreading ridges are the driving force for plate motion: the zigzag pattern seems much more likely to be an effect of the break-up of a strong lithosphere (e.g. see Fig. 6.14).

Lead isotope evidence

The lead isotope compositions calculated for basalts from the Tristan da Cunha and Ascension hot-spots show a two-stage mixing: one at 4.5 Ga (when the mantle formed) and a second at 1.8 Ga, together with possible further changes in the last few Ma. Morgan takes this to indicate that the material making up these volcanic islands was last near the surface 1.8 Ga ago, and that 1.8 Ga is the 'cycle time' required for a particle to sink slowly through the mantle and then rise again in a plume. This gives a rate of upwelling of 500 km^3 per year. If the total volume of the mantle (10^{12}km^3) is divided by this rate, a figure of 2 Ga results. He concludes from this that the whole mantle must be involved in the convective overturn, and therefore that the plumes must extend right down to the core–mantle boundary.

The role of trenches

Plates with long trench boundaries in general move much faster (by a factor of about x4) than plates with no, or very short, trench boundaries (Fig. 10.7). This is taken to indicate that slab-pull is a more effective driving mechanism for plate motion than ridge push. Tensile stresses at trenches are estimated at a few hundred bars (however, there is considerable variation).

Plumes drive the plates

Morgan concludes as follows:

> '…the mid-ocean positions of most of the plumes and the land evidence of plume activity prior to continental break-up suggest that the plumes produce the stresses which drive the plates apart.'

How do plumes fit into mantle convection?

McKenzie had concluded in 1969 that the cold descending slabs must control the positions of the descending limbs of any convection cell. Consequently, if the subduction zone moves, as it does with trench roll-back, the sinking current must move with it. Moreover, when continents collide, the site of a sinking current must eventually be replaced with a new sinking current somewhere

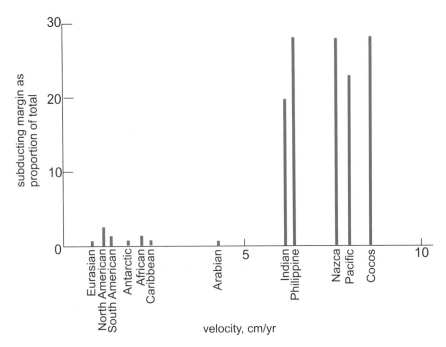

Figure 10.7 Proportion of plate margin connected to subducting slab compared to absolute plate velocity. Plates with long subducting margins are significantly faster. After Forsyth & Uyeda, 1975.

else in order to maintain the overall convection system. This implies that convection cells must continually change as the global plate-tectonic system changes: this is inevitable because of the constraints imposed by the strength and geometry of the plates – as shown, for example, by the lateral movement of ridges away from their original mantle source.

A good example of the long-term movement of the ocean ridge system is the Tristan da Cunha hot-spot, mentioned earlier, that now lies several hundred kilometres east of the present position of the mid-Atlantic ridge, the latter having moved westwards away from its original position over the hot-spot. Another obvious example is the eastern Pacific, where the original ridge system has been over-ridden by the westwards movement of the Americas plate. As mentioned above, the Antarctic plate is almost completely surrounded by ridges and is consequently growing in size over time, owing to the lack of subduction zones. These ridges must therefore also be moving away from their original sources. However, the link between falling currents and subduction zones may be more long-lasting, since the sinking slab may act as an anchor to the subduction zone, preventing it from too much lateral movement away from its mantle root.

It became clear that the whole-mantle convection cells envisaged by Holmes and Runcorn (see chapter 4) were inadequate to describe a convection system governed in its upper levels by the transient nature of the plate boundary system, and many geophysicists have preferred a model characterized by relatively small-scale 'eddies' in the upper part of the mantle and a larger-scale convective circulation involving the whole mantle, as envisaged by Morgan. The seismic discontinuity at $c.650$ km depth was thought to represent a possible barrier to through-going convection. Hot-spots are associated with topographic swells of around 1 km amplitude and also with positive geoid anomalies with amplitudes of abut 10 m. Both these structures are elongated in the direction of plate motion, which may indicate a convective pattern with size and spacing different in scale from that associated with the major circulation of the plate movements.

In Jason Morgan's plume model, two largely independent convective processes in the mantle were proposed: the broad convective flow associated with plate tectonics and driven primarily by sinking slabs of cold lithosphere plates back into the asthenosphere, and mantle plumes which carry heat upwards in narrow, rising columns, driven by heat exchange across the core–mantle boundary. The latter type of convection was thought by some to be independent of plate motion, but Morgan clearly believed that continental break-up, and thus the initiation of major changes in plate tectonics, was initiated by plumes.

The last major re-organization of the global convection system was probably associated with the amalgamation of Pangaea at the beginning of the Mesozoic Period. Pangaea then formed around 40% of the Earth's surface, and was almost entirely surrounded by subduction zones. These would have imposed a relatively simple pattern of curved sheet-like sinking currents around the Pangaea margin, and an equally simple pattern of ridges in the ancestral Pacific Ocean, radiating perhaps from a central plume. Plumes may have played a major role in disrupting this global system.

Later developments

The work of Wilson and Morgan stimulated extensive research activity directed towards the role of plumes in mantle convection; this research was focused particularly on four main lines of enquiry: laboratory experiments, the interpretation of geoid anomalies, seismic wave analysis, and geochemical/ petrological models.

Laboratory experiments

Several geophysicists have attempted to model mantle convection by means of laboratory simulations. A good example is the series of experiments carried out by John Elder and described in his 1976 book entitled *The bowels of the Earth*.

Elder's experiments used a container of hot oil allowed to cool from above. He found that a cooled, highly viscous sub-layer was formed at the surface, which acted as a buffer between a vigorously convecting interior and the cool surface, held at a constant temperature. The interior flow tended to continually thin the sub-layer until it overturned, which enabled convection to continue. A series of eddies was formed, on a scale similar to the depth of the sub-layer, which entrained cool material from the surface and brought hot material closer to it. Eventually a cool blob of fluid fell out of the sub-layer into the interior, thereby disrupting the integrity of the sub-layer.

Elder also showed that the edge of a floating surface slab (simulating a continent) induces an asymmetrical eddy that propels the slab sideways, and produces an upwelling current behind the leading edge of the slab, thus simulating volcanic arc production. Other experiments have simulated the production of plumes by heating from below.

This behaviour could thus present an analogy of lithospheric behaviour and the subduction process. There are several conclusions that could be drawn from these experiments.

1. The behaviour of the cooled sub-layer (i.e. the lithosphere) is critical to the convection process.
2. Plume-like structures are the main vehicle for transferring heat upwards.
3. Continents generate their own system of small-scale currents that produce lateral forces and volcanic activity.
4. Descending cool currents (i.e. subduction zones) exert a major control on the convective system.
5. Continental plates do not ride passively on the back of horizontal mantle flow (i.e. plates are not driven by mantle drag).
6. Plate movement is not primarily controlled by upwelling currents.

The problem with these kinds of experiments is that, in order to reproduce processes in the Earth that take millions of years, the physical properties of the material used (in Elder's case the oil), and the size and shape of the container, have to be realistically scaled. Thus to reduce the timescale to minutes or hours, the viscosity of the material has to be reduced to match, speeding up the observed processes by about $\times 10^{14}$. However, this involves assumptions about the actual physical state of the mantle (i.e. its density, temperature and composition etc.). Not enough information exists about these factors to be able to judge whether the experiments offer a reliable guide to actual mantle behaviour. Moreover, the behaviour of a single material such as oil cannot reproduce the complex interchange of chemical components that is known to take place within the petrological Earth.

Mathematical models simulating convection and plume behaviour have also been produced, but these suffer from the same disadvantages as the laboratory experiments. The real Earth system is extremely complex, and many of its characteristics and properties are poorly known, so that any model of convection or plume behaviour can only be a guide to what might happen rather than being a representation of the actual system.

The evidence from the geoid

The geoid describes the shape of the Earth's surface as represented by mean sea level (or the equivalent level on land, if the sea were allowed to flow there) and is represented by the gravitational field, which can be measured from satellites by sensitive gravimeters. This shows that the Earth is not spherical but is flattened at the poles and also exhibits large-scale bulges (positive anomalies) and depressions (negative anomalies) up to 100 m in size and around 4,000 km across (Fig. 10.8). These large-scale effects are superimposed on the state of general gravitational equilibrium where much of the large topographic differences between oceanic and continental crust, and between mountain ranges and coastal plains, are cancelled out.

Because of their size, the large-wavelength anomalies are assumed to reflect deep-seated mantle heterogeneities rather than lithospheric effects. In terms of mantle convection, the positive anomalies could correspond to cold dense downward currents, or aggregations of old dense subducted material, producing a mass excess within the deep mantle, whereas the negative anomalies could

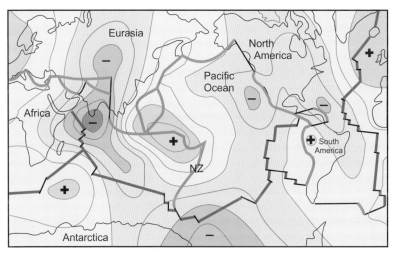

Figure 10.8 Long-wavelength anomaly pattern of the geoid, contoured at 20 m intervals, compared to the plate-boundary network. Ridges, green; convergent boundaries, orange; transform faults, black. Note that positive areas (blue) are located over the western Pacific rim and in the North Atlantic, whereas the negative areas (red) are in the Indian Ocean, Central Asia, the Eastern Pacific and the Antarctic Ocean. Adapted from McKenzie, 1969.

correspond to a mass deficiency at depth, produced by a warm, less dense, mantle plume. It has been suggested that these density anomalies may represent swells and depressions on the core–mantle boundary: the swells associated with the sinking slabs and the depressions with rising warm currents or plumes. The major arc–trench systems of the Western Pacific and the Andes all lie within large-scale positive anomalies, whereas the large negative anomalies occur in the Indian Ocean, the eastern Pacific–Caribbean region, and the Antarctic Ocean south of New Zealand (Fig. 10.8).

The geoid also exhibits shorter-wavelength anomalies of the order of 10 m in size and with half-wavelengths of the order of 1,000–1,500 km across. These smaller-scale anomalies are measured with reference to the local large-wavelength geoid and may either accentuate or partly offset the regional anomaly. These types of anomaly are more likely to be caused by mass imbalances within the crust, or to reflect an upper-mantle circulation pattern that is smaller in scale than the large-scale circulation associated with the major plates. Morgan suggested that positive anomalies of this type were caused by mass excesses associated with mantle plumes. In the case of Hawaii, for example, both the positive gravity anomalies and the topographic swells associated with them are elongated in the direction of the linear seamount chain. The largest short-wavelength negative anomalies are associated with deep trenches, whereas the wide, fast-spreading ridges have very small anomalies, because their topography is largely compensated.

The interpretation of the geoid anomalies has proved difficult and controversial, and although there is now a much more accurate map of the geoid, there is as yet no generally agreed interpretation linking the anomaly pattern to a specific mantle structure. The solution to the problem has to depend on other types of evidence, such as seismic tomography or geochemistry.

Seismic tomography

The basic structure of the Earth's interior had been known from seismic wave analysis for many decades but the development of more powerful computers enabled more sophisticated analysis of seismic wave patterns to be carried out by means of the process of 'seismic (wave) tomography'. This requires the analysis of large quantities of earthquake arrival data in order to define areas of anomalous velocity or attenuation (i.e. 'Q' values). These are analysed through a series of superimposed two-dimensional layers or shells in order to build up a three-dimensional picture of mantle properties.

Attenuation tomography is sensitive to temperature variation and can distinguish thermal features such as subduction zones or hot-spots (see Fig. 9.8). The results of this method have shown that large-scale anomalies can be detected that seem to represent cold, dense bodies such as subducted slabs or warm, less dense bodies such as 'super-plumes' on a scale of hundreds of kilometres. However, the degree of resolution of the method is such that bodies with smaller radii or thickness, such as narrow plumes, would not be imaged.

Super-plumes have also been detected in the form of large regions of anomalously low shear-wave velocity, believed to indicate thermal and compositional differences. For example, a well-defined low-velocity body was detected beneath the Yellowstone hot-spot, extending between 30 km and 250 km in depth, followed by a less distinct body between 250 km and 650 km depth, and inclined at 60° to WNW. The Hawaii hot-spot was found to lie above a similar body 500–600 km wide and 2000 km deep. Subduction zones have also been imaged by the same method below both the Farallon plate and the northern edge of the Indian plate.

Thus seismic tomography has confirmed that hot-spot concentrations correspond to low-velocity, low-Q (and presumably hot) regions of the mantle and thus would seem to validate the plume concept.

A geochemical/petrological model

It had been accepted for several decades that the mantle was essentially composed of varieties of peridotite consisting of a mixture of iron-magnesium silicates, plus minor constituents that varied depending on how 'fresh' or 'fertile' the material was. The fresh peridotite, often termed 'pyrolite', which is capable of melting to produce basalt, leaves a denser form of peridotite (harzburgite) as a residue after some basaltic melt has been extracted. This residual peridotite is enriched in certain constituents, such as the relatively immobile elements titanium, niobium and tantalum. Also, because the magmas produced as a result of the partial melting are enriched in the more mobile constituents such as the alkalis potassium and rubidium, and the so-called incompatible elements such as strontium, uranium and lead, the residual peridotite is relatively impoverished in these constituents.

Ocean-island basalts and volcanic seamounts far from any subduction zone are typically higher in iron/magnesium and titanium/magnesium ratios, lower in alumina, and higher in the incompatible trace elements than the mid-ocean ridge basalts (MORB), as are certain basalts from continental interiors, such as the Yellowstone hot-spot. In contrast, the MORB basalts are relatively depleted in the alkalis and incompatible elements.

In an interesting contribution to the plumes debate in 1982, experimental petrologist Ted Ringwood of the Australian National University addressed the origin of plumes from the point of view of the fate of the individual components of the oceanic lithosphere as they descend through the mantle. He pointed out that there are three separate components, each of which would be expected to behave differently as they descend and encounter higher temperatures and pressures:

1. the basaltic component of the crust, which would normally be expected to have been converted first to amphibolite and eventually to eclogite;
2. the harzburgite layer of 'exhausted' pyrolite immediately beneath it; and
3. a layer consisting of depleted pyrolite from which only some basaltic melt has been extracted.

During subduction, water-soluble constituents are extracted from the basaltic component of the sinking slab, leaving it enriched in the more immobile elements. Then, as the slab continues downwards, the lowermost layer of relatively ductile pyrolite would be stripped off and resorbed back into the surrounding mantle. This material would be depleted in highly incompatible elements such as niodymium, strontium and lead, leading to progressive depletion of these within the upper-mantle source region for MORB basalts. The remainder of the slab, which continues downwards, will be denser than the surrounding mantle and would sink to the base of the upper mantle at the 650 km boundary. Here, Ringwood envisages a mass of dense slab material, which he terms a 'megalith', forming over a period of time lasting perhaps 1–2 Ga during which it gradually warms up, the basaltic component partially melts and contaminates the adjoining harzburgite, rendering it more 'fertile'. The remaining blocks of dense ex-basaltic material (now eclogitic?) would remain and possibly fall through the lower mantle. The newly fertile ex-harzburgitic element would now be buoyant and could rise up through the upper mantle in the form of plumes to produce 'hot-spot-type' alkaline volcanicity enriched in incompatible elements.

In this model, the MORB volcanicity is derived from a pyrolite source region in the upper mantle that has experienced the episodic extraction of small amounts of alkalic magmas and is consequently depleted in incompatible elements. However, as megalith material is transferred to the lower mantle over periods of the order of 1 Ga or more, it is envisaged that, in order to sustain the convection system, the upper mantle would be periodically refreshed by upwelling plumes of 'primitive' pyrolite from the lower mantle.

Thus, in Ringwood's model, two types of plume-type diapiric bodies are visualized: those confined to the upper mantle, providing the source for the hot-spots; and larger bodies ('super-plumes') arising from the core–mantle boundary, which are responsible for periodic replenishment of the upper-mantle source region for the MORB volcanics. The super-plumes would be responsible for the long-wavelength geoid anomalies, and the upper-mantle plumes for the smaller-scale anomalies.

Subsequent models have suggested that super-plumes from the core–mantle boundary could have been contaminated by accumulations of ex-basaltic material lying at the base of the mantle, and it is these that are responsible for the major hot-spots associated with continental break-up.

Distinguishing between whole-mantle and two-layer convection models requires a knowledge of the nature of the 650 km discontinuity. If this discontinuity represents a phase transformation between Mg_2SiO_4 with a spinel-type structure and $MgSiO_3$ with a perovskite-type structure plus $(Mg, Fe)O$ magnesio-wüstite, as has been suggested, then this would permit whole-mantle convection to occur freely between the upper and lower mantle, since the transition would be gradual. However, if the discontinuity represents a change in composition from, say, Mg_2SiO_4 olivine plus spinel to $MgSiO_3$ 'perovskite', the transition would be expected to be sharper, and the convective systems of the upper and lower mantle would be essentially independent and separated by the 650 km boundary layer. Not enough is known about the physical properties of the mantle to discriminate conclusively between these two possibilities, although it has been argued that the discontinuity is too sharp to be caused solely by an isochemical transition.

Convection within the upper mantle would have to be very efficient in order to provide the degree of mixing required to explain the high degree of chemical and isotopic homogeneity displayed by the MORB source region. If, on the other hand, replenishment of the MORB source region were achieved by plumes of geochemically primitive pyrolite from the lower mantle, pushing straight through the boundary layer, the high degree of homogeneity is easier to explain. Ringwood's model also explains the variability of the intraplate basalts compared with those of the MORB, since many of the isotopic characteristics of the former would have been derived originally from former basaltic ocean crust that had been subducted and stored in the mantle – perhaps for more than 1 Ga.

Postscript

The mantle plume concept, introduced by Tuzo Wilson in the 1960s and elaborated by Jason Morgan in the 1970s after plate-tectonic theory had been accepted by the geological community, stimulated a widespread debate that transformed ideas on mantle convection. No generally accepted model has yet appeared, although the basic idea of plumes arising from the base of either the upper or lower mantle, as integral parts of a convection system, does seem to have become established. There can be no doubt however that Wilson's original idea had made a critical contribution. The debate continues!

11

Sequence stratigraphy

Two American geologists, Laurence Sloss and Peter Vail, were primarily responsible for introducing the concept of sequence stratigraphy to the geological world. This concept arose out of the work of Sloss, together with H.E. Wheeler and others in the 1940s and 1950s, in the investigation of the cratonic cover of the North American continental interior. The basic idea was to define packages of sedimentary strata in terms of their unconformable (i.e. erosional) boundaries rather than their lithological similarities. Whereas it was well known that a particular sedimentary lithology, such as sandstone or shale, was likely to be diachronous, Sloss argued that an unconformity must represent a time-line, or at least a recognizable time interval, and that sequences of strata bounded by unconformities must represent distinct chrono-stratigraphic units that could in theory be correlated on a continent-wide, or even a worldwide, basis. In 1963 he defined the stratigraphic sequence as:

> 'a rock-stratigraphic unit of higher rank than group, megagroup or supergroup, traceable over major areas of a continent and bounded by unconformities of inter-regional scope.'

However, little use was made by stratigraphers of the sequence stratigraphy concept until a practical application became evident in the later 1960s and 1970s, when a group of Exxon petroleum exploration geologists led by Peter Vail, a former postgraduate student of Laurence Sloss, applied it to the interpretation of seismic reflection data. The method is described in 1977 in a much-referenced memoir of the American Association of Petroleum Geologists edited by C.E. Payton. This work transformed the way in which both stratigraphers and sedimentologists approached stratigraphic sequences and revolutionized the methods used by exploration geologists to interpret their data.

Historical background

Stratigraphy

The introduction of stratigraphy as part of the earth sciences is credited to the English canal engineer William Smith (1769–1839), who observed that the rock layers within a region were arranged in a predictable pattern, could always be found in the same relative positions, and contained the same types of fossil.

His geological map in 1799 of an area around Bath in Somerset (England) is believed to be the first of its kind. Since then, geologists have classified sedimentary rock units in terms of 'formations' based on lithological similarity, such as 'the Chalk', 'the Oxford Clay', or 'the Wenlock Limestone', which are essentially local units that may or may not be correlatable regionally. Charles Lyell's *Elements of Geology*, first published in 1838, defines a formation as:

'any assemblage of rocks which have some character in common, whether of origin, age or composition.'

Eventually stratigraphic classification became more formalized into a hierarchical pattern where formations were subdivided into 'members' each composed of individual beds. In turn, formations themselves were combined into groups and then megagroups or supergroups – all based on lithological similarity.

However useful the lithostratigraphic system may have been in constructing regional or even national geological maps, it was of limited value in interpreting the regional or global processes that ultimately explained the stratigraphic record. The lithostratigraphic system had first to be integrated into a chronological (i.e. 'chrono-stratigraphic') framework, based initially on fossils and more recently also on increasingly accurate radiometric dating, in order for regional and global correlations to be made and hypotheses advanced as to causal processes.

Sedimentology as a separate branch of the earth sciences is concerned with the processes that are responsible for assemblages or packages of strata as much as, or probably more than, the origin and description of individual sediment types. It was well known that, for example, individual beds of sandstone would be likely to grade laterally into finer-grained siltstone, then mudstone, and ultimately die out. It followed that a stratigraphic unit consisting of sandstone overlain by siltstone then mudstone should be considered as a single sedimentation 'event', and that an overlying unit beginning with a coarse sandstone would represent a separate event. If stratigraphy were to be interpretable in terms of events, then the important unit was the sequence of sediments, and the time-line represented by the break (unconformity or disconformity) separating two sequences was thus an important, potentially correlatable, structure representing an event – of sedimentation, erosion etc. This insight was the basis of sequence stratigraphy.

Sequence stratigraphy

The Wikipedia definition of sequence stratigraphy is a good starting point:

'Sequence stratigraphy is a branch of geology that attempts to subdivide and link sedimentary deposits and unconformity-bound units on a variety of scales and explain these stratigraphic units in terms of variations in sediment supply and variations in accommodation space (often associated with changes in relative sea level). The essence of the method is mapping of strata based on identification of surfaces which are assumed to represent time lines (e.g. subaerial unconformities, maximum flooding surfaces), and therefore placing stratigraphy in [a] chronostratigraphic framework.'

The basic model

This employs the concept that truncated bedding surfaces define either sedimentary or erosional 'events', which could be interpreted in terms of processes, such as transgression or regression of sea level or lake level, etc. The important unit in stratigraphy is thus the sequence of genetically related sedimentary deposits, and the sequence boundaries are the most significant surfaces. These boundaries are either unconformities or the conformable boundaries (disconformities) that correlate with them. Such boundaries are formed as a result of a fall in sea level. This causes erosion of part or all of the previously deposited sequence. Figure 11.1 displays an idealized system consisting of three depositional sequences separated by unconformities. Note that the truncated beds of the top of the lower sequences (i.e. 5–10 and 18–19 on Fig. 11.1) are the result of erosion, whereas those of the base of the upper sequences (11–15 and 21–23) result from non-deposition. A complete chronology can only be obtained by identifying the highest un-truncated unit of a lower sequence (i.e. 10 and 19 respectively) and the lowest unit of the overlying sequence (11 and

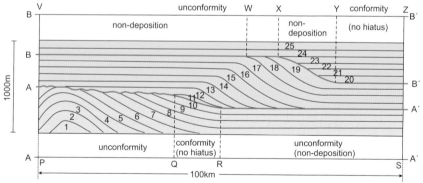

Figure 11.1 Stratigraphic geometry of an idealized depositional sequence (orange beds) bounded by two unconformities A–A´ and B–B´. Note that the base B–B´ of the upper set of beds (yellow) is unconformable from V to W due to non-deposition of beds 18–23 although apparently conformable. Similarly, the base A–A´ of the middle sequence (orange) is unconformable from R to S, although apparently conformable due to non-deposition of beds 11–15. The only obvious angular unconformities due to erosion are along sections P–Q on A–A´ and X–Y on B–B´. After Mitchum et al., 1997.

20). The latter may only be discovered, for example, at the bottom of a valley incised into the lower sequence.

The Wheeler diagram

An alternative way of displaying the information on Figure 11.1 is a chrono-stratigraphic chart, where the vertical axis represents time, rather than depth, as in a traditional stratigraphic chart. Such a diagram was introduced by H.E. Wheeler in 1958 and Figure 11.2 shows the data of Figure 11.1 displayed on a Wheeler chart. The advantage of this method is that the extent of the periods of erosion or non-deposition can be displayed in their true chronological context, giving a better picture of the processes involved. Thus the unconformity A–Á can be clearly seen as erosional between P and Q, cutting through beds 6 to 10, but also non-depositional in respect of beds 11 and 12. Similarly, this unconformity between R and S is also progressively non-depositional with respect to beds 12 to 15. The only truly complete sequence is between Q and R.

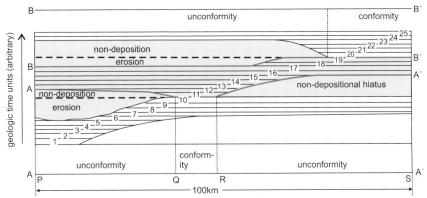

Figure 11.2 Chrono-stratigraphic chart (Wheeler diagram) of the same stratigraphic succession as Figure 11.1, but showing the breaks in the succession caused by depositional and erosional periods in grey; the sequence between A–A´ and B–B´ is shown in yellow. After Mitchum *et al.*, 1997.

The contribution of Laurence Sloss

Laurence L. Sloss (1913–1996) was born in California and graduated at Stanford and Chicago Universities. He was appointed to the faculty at Northwestern University, Illinois, in 1947, ultimately becoming William Deering Professor of geological sciences there in 1981. While at Chicago, he was influenced by two already distinguished stratigraphers, Francis Pettijohn and Bill Krumbein, and followed the latter to Northwestern. Collaboration with Krumbein resulted in their well-known textbook *Stratigraphy and sedimentation* published in 1951.

Sloss is credited with revolutionizing the science of stratigraphy and was awarded the Penrose medal of the Geological Society of America in 1986 in recognition of his pioneering work on sequence stratigraphy. Sloss's defining contribution on sequence stratigraphy was published in 1963, with the title: *Sequences in the cratonic interior of North America.*

It was from this 1963 paper that the concept of stratigraphic sequences first obtained widespread attention, but Sloss himself attributes the initial concept to a paper jointly authored by himself, along with his colleagues W.C. Krumbein and E.C. Dapples, which was presented to the Annual Meeting of the Geological Society of America in 1948. However, it is the 1963 paper that is usually cited, and it is here that the concept and its applications to the North American craton are reviewed in detail.

Sloss notes that inter-regional correlations within the North American continental interior had suffered from the setting up of separate and markedly different stratigraphic sequences, using local nomenclature, on the opposite sides of once-continuous basins, which have resulted in complex and inconsistent correlation charts:

> 'the records of basins which are now separated by areas of post-depositional erosion but which show no evidence of having been separated at the time of deposition thus appear to bear no close relationship to one another in terms of the time-stratigraphic correlation of the strata involved. As a result, inter-regional relationships form an apparently chaotic pattern.'

Utilizing the widespread drilling records in the sedimentary basins, Sloss concluded that there was a 'fundamental homogeneity' in the stratigraphy of the cratonic interior of North America in spite of the regional differences of nomenclature and description. Because much of the problem of correlation had stemmed from the failure to distinguish between local and regional unconformities, Sloss recognizes three basic types: biostratigraphic hiatus (i.e. a break in the fossil record), local unconformity and regional unconformity. He notes that many examples of recorded biostratigraphic gaps occur within continuous and gradational sedimentary cycles, and that these cannot represent genuine unconformities, but must have a biostratigraphic cause, perhaps reflecting sedimentary facies changes (e.g. from mudstone to sandstone). Other examples of biostratigraphic breaks do correspond to physical breaks in the succession although there is apparent conformity: i.e. the beds above and below are parallel (the term 'disconformity' is sometimes used for such categories). Figures 11.1 and 11.2 show several examples of these, some due to erosion without any tilting of the underlying strata (e.g. the section V–W in Figure 11.1) and others due to non-deposition, where the base of the upper beds gradually transgresses over a contemporary landmass (e.g. the section X–Y in Figure 11.1). Unconformities are common along basin margins due to tectonic processes related to the orogenies taking place along the western and eastern sides of the craton, many of them recording abrupt angular discordances, but few can be traced far into the continental interior and must be regarded as of local significance only.

Figure 11.3 is a chrono-stratigraphic table illustrating Sloss's concept of the stratigraphic sequence as applied to the North American craton. Sloss concludes that only six of the unconformities recognizable in the cratonic interior can be shown to be truly inter-regional in scope, although none of them exhibit obvious characteristics that distinguish them from the many others that are not. Each of these extends across the sedimentary basins of the craton interior and passes into the miogeosynclinal basins at the craton margin, where they either pass into a conformable sequence or are unrecognizable among the many local unconformities. The sequences separated by these six unconformities are the basis of Sloss's concept of the stratigraphic sequence. In terms of continent-wide events, their bases represent major transgressive phases during which

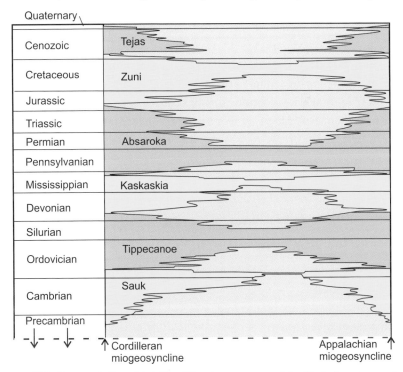

Figure 11.3 Chrono-stratigraphic table of the six sequences of the North American craton according to Sloss. The names of the sequences (Sauk, Tippecanoe etc.) were introduced by Sloss to distinguish them from the conventional stratigraphic names. Areas of sedimentation shown in yellow or blue; grey areas represent non-deposition. After Sloss, 1963.

sedimentation begins at the craton margins and spreads gradually across the stable continental interior.

Sloss concludes by noting that none of the six sequences corresponds to any of the Systems (Periods) of the international Stratigraphic Column, leading to the use of terms such as 'Cambro-Ordovician', 'Permo-Carboniferous' etc. and reflecting the fact that the system boundaries were first erected in Europe and are not a natural fit for North America, where the 'natural' groupings of strata cross the system boundaries.

In his 1963 paper, Sloss applies the stratigraphic sequence concept only to major continent-wide units, however the usage has subsequently broadened to include any unconformity-bound unit of whatever scale, as will be seen in its application in seismic stratigraphy.

Seismic stratigraphy

Seismic stratigraphy depends on the assumption that seismic reflection surfaces correspond to (usually) parallel bedding surfaces, which can be regarded as time markers, enabling correlation to be made over wide areas. Conventional outcrop analysis had previously been reliant on the correlation of lithofacies (i.e. sandstone with sandstone, shale with shale, etc.) which, as was well known, were typically diachronous (Fig. 11.4). Any one sandstone body, for example, would normally grade laterally into finer-grained siltstone then mudstone and eventually die out as the sediment supply ceased. To be of any value in the analysis of the thee-dimensional architecture of a particular sedimentary basin – critical in petroleum exploration – a sedimentary sequence must be integrated into a chronostratigraphic framework so that it can be correlated regionally.

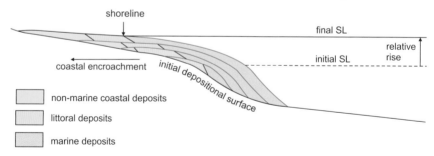

Figure 11.4 Diachroneity of lithofacies during a relative rise in sea level. The sedimentary facies vary laterally from non-marine coastal deposits (e.g. estuarine sands) to littoral (i.e. intertidal deposits) to marine shales, so that individual lithofacies units cross the isochrons (time lines) shown in red.

The methodology of, and background to, seismic stratigraphy are comprehensively explained in the book by Andrew Miall published in 1997 entitled *The geology of stratigraphic sequences*. According to Miall, seismic stratigraphy:

'... permitted two major practical developments in basin analysis, the ability to define complex basin architecture in considerable detail, and the ability to recognize, map and correlate unconformities over great distances.'

The application of sequence stratigraphy to petroleum exploration was carried out by the Exxon team of geologists and geophysicists led by Peter Vail, essentially as a tool in finding petroleum deposits, but in doing so a valuable new way of interpreting sedimentary sequences was opened up. Access to a large worldwide database of well records and bio-stratigraphic information in addition to the seismic records resulted in the famous and much referenced Memoir 26 of the American Association of Petroleum Geologists (AAPG), published in 1977.

Peter Vail and the Exxon team

Peter Vail (1930–) obtained his master's degree and doctorate at Northwestern University, Illinois, under the supervision of Laurence Sloss. On graduating in 1956, he joined the Exxon Petroleum Corporation (now ExxonMobil) where he remained for thirty years. He left in 1986 for a teaching post at Rice University, Houston, where he is currently Maurice Ewing Professor Emeritus in the Department of Earth Sciences. He has received numerous medals and honours for his pioneering work on seismic stratigraphy, including the Benjamin Franklin Medal in 2005.

The essence of the seismic stratigraphy concept and methodology is explained in the aforementioned AAPG Memoir 26. The first part of the memoir deals with the methodology and limitations of seismic interpretation. The second part is entitled *Seismic stratigraphy and global changes in sea level* and includes 11 separate papers authored by geologists of the Exxon Production Research Corporation led by Peter Vail.

The basis of the method is explained in an introductory article by Vail, jointly with R.M. Mitchum, Jr.:

'primary seismic reflections parallel stratal surfaces and unconformities. Whereas all the rocks above a stratal or unconformity surface are younger than those below it, the resulting seismic section is a record of the chronostratigraphic (time-stratigraphic) depositional and structural patterns and not a record of the time-transgressive lithostratigraphy (rock-stratigraphy).'

It follows that:

'Because seismic reflections follow chronostratigraphic correla-tions, it is not only possible to interpret post-depositional structural deformation, but also it is possible to make the following types of stratigraphic interpretations from the geometry of seismic reflection correlation patterns: (1) geologic time correlations, (2) definition of genetic depositional units, (3) thickness and depositional environment of genetic units, (4) palaeobathymetry, (5) burial history, relief topography on unconformities, and (6) palaeogeography and geologic history when combined with geologic data.'

The key concept here is the recognition that the surfaces defined by seismic stratigraphy are time lines that are potentially correlatable, in distinction to lithologic units that are not. The authors adapt the approach initiated by Sloss of identifying stratigraphic units composed of relatively conformable successions, or depositional sequences, the upper and lower boundaries of which are unconformities or their correlatable conformable horizons. Thus they state:

'the time interval represented by strata of a given sequence may differ from place to place but the range is confined to synchronous limits marked by ages of the sequence boundaries where they became conformities.'

Terminology

The methodology is described in detail in the immediately following article by Mitchum, Vail and Thompson (1977a). The first step in understanding and interpreting the geometric patterns produced by seismic records was the development of a suitable terminology in order to describe the shape and character of stratigraphic surfaces and their terminations. Here the authors redefine various stratigraphic terms, and introduce a number of new ones. This enabled the data to be interpreted in terms of processes, such as transgression, regression and erosion, which could be linked in turn with rises and falls of relative sea level, changes in sediment supply, or tectonic movements involving uplift or depression. The more important of these terms are defined as follows (not all of them still routinely used, however):

- 'A *sechron* is the total interval of time during which a sequence is deposited.' It is measured between the conformities at the top and base of the sequence (which may be in different places).
- 'A *hiatus* is the total interval of time that is not represented by strata at a specific position along a stratigraphic surface.' Hiatuses may be attributable either to erosion or non-deposition or both (e.g. see the grey areas in Fig. 11.2). The magnitude of a hiatus is of the same order as that of the whole sequence – i.e. where a sequence represents a few Ma in time span, its absence suggests a similar magnitude.

Clinoforms are the individual bedding surfaces, and are of two basic types: either sigmoidal in shape, with a gentle dip at their proximal end and a steeper

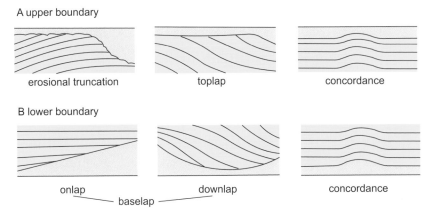

A upper boundary

erosional truncation   toplap   concordance

B lower boundary

onlap   downlap   concordance
baselap

Figure 11.5 Explanation of terms for various types of boundaries to stratal sequences. **A** upper boundaries; **B** lower boundaries. In each case, the grey area represents the adjoining sequence. See text and Figure 11.8 for further explanation. After Mitchum *et al.*, 1997.

dip at their distal ends, where they are truncated at the downlap surface; or oblique, where they are cut off both above and below (e.g. see Fig. 11.6). Bedding dips are typically less than 1° in the sigmoidal clinoforms but may be up to 10° in the oblique clinoforms.

The geometry of the discordance at a sequence boundary is the main criterion used to determine the nature of the boundary, i.e. whether it results from erosion or non-deposition. The direction of progressive terminations from older to younger strata above an unconformity corresponds to the direction of increasing non-depositional hiatus along the unconformity. The following terms are used to define the nature of discordances, some of which are illustrated in Figure 11.5.

- '*Lapout* is the lateral termination of a stratum at its original depositional limit.'
- '*Truncation* is the lateral termination of a stratum as a result of being cut off from its original depositional limit.'
- '*Baselap* is lapout at the lower boundary of a depositional sequence.'
- '*Onlap* is a type of baselap in which an initially horizontal stratum laps out against an intially inclined surface, or where an initially inclined surface laps out against a surface with a greater inclination.'
- '*Downlap* is baselap in which an initially inclined stratum terminates downdip against an initially horizontal or inclined surface.'

Onlap and downlap are indicators of non-depositional hiatuses rather than erosional ones.

- '*Toplap* is lapout at the upper boundary of a depositional sequence.' Initially inclined strata may show this – the lateral terminations may taper and approach the upper boundary asymptotically. Toplap is evidence of a non-depositional hiatus and occurs when the depositional base level (e.g. sea level) is too low to permit strata to be deposited further up-dip.

- '*Erosional truncation* is the lateral termination of a stratum by erosion and occurs at the upper boundary of a depositional sequence'.
- '*Structural truncation* is the lateral termination of a stratum by structural disruption'. The disruption may have a number of causes, such as faulting, gravity sliding, salt flowage or igneous intrusion.

Examples of seismic reflection patterns illustrating these types of discordance are shown in Figures 11.6 (top-discordant patterns) and 11.7 (base-discordant patterns).

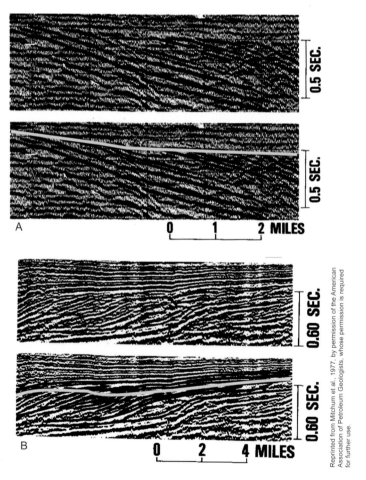

Figure 11.6 Top-discordant seismic reflection patterns: **A** erosional truncation; **B** toplap. The yellow line shows the unconformity in A and the toplap surfaces in B. The lower photographs show the interpretation. From Mitchum *et al.*, 1977, with permission.

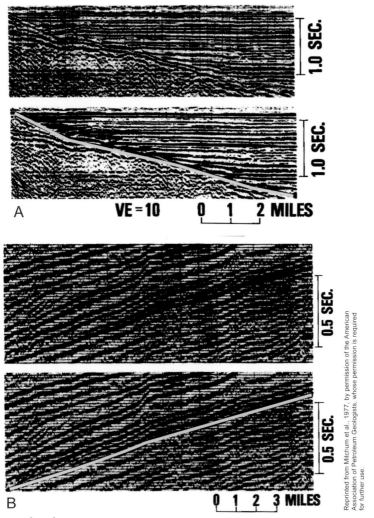

Figure 11.7 Base-discordant seismic reflection patterns: **A** onlap; **B** downlap. The lower of each pair is interpreted. In B, the downlap surface is indicated by the yellow line. The downlapping strata above it make a shallow angle with it. From Mitchum *et al.*, 1977 (part 6), with permission.

Reprinted from Mitchum et al., 1977, by permission of the American Association of Petroleum Geologists, whose permission is required for further use.

Systems tracts

These are linked contemporaneous deposition systems, each defined by the stratal geometry at their bounding surface, their position within the sequence, and their internal stacking pattern (Fig. 11.8). They are interpreted as a function of the interaction between 'eustasy' (i.e. global changes of sea level), sediment supply and tectonics. There are three principal systems tracts: 'lowstand', 'transgressive' and 'highstand'. The lowstand systems tract (LST) develops on the continental slope and basin floor at times of low relative sea level; the

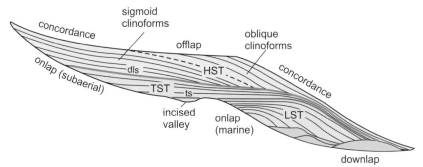

Figure 11.8 The basic sequence-stratigraphic model illustrating the different types of surface and their terminology. The basic sequence is divided into two sections: the lower (pink) consists of an upper transgressive systems tract (TST) and a lower, lowstand systems tract (LST); these onlap on a basal unconformity, marine at its lower end and subaerial at its upper; the upper section (yellow) consists of a regressive highstand systems tract (HST) exhibiting offlap. Both sections exhibit downlap at the seaward end (i.e. successive bedding surfaces are cut off at the base). The discordance separating the two sections (blue line) is a downlap surface (dls) representing the beginning of a marine regression. Individual bedding surfaces (clinoforms) are either sigmoidal (with an upper concordant section) or oblique (truncated at both base and top). Note the incised valley at the base of the lower section. The green line (ts) is a transgressive surface. The orange units at the termination of the lowstand surfaces represent successive submarine fan deposits. After Miall, 1997.

transgressive systems tract (TST), develops at the top of the LST, and marks a marine transgression characterized by shelf or basin-floor facies such as marine shales or platform carbonates. The top of the TST corresponds to the 'maximum flooding surface' (MFS); this is a downlap surface above which is the highstand systems tract (HST); this forms the upper part of the stratigraphic sequence, and typically consists of shelf or non-marine deposits, which may be partially eroded by a transgressive surface marking the base of the succeeding sequence.

The basic seismic sequence model represented in Figure 11.8 illustrates the more important of these descriptive terms. The model is divided into two sections, a lower and an upper. The base of the lower section exhibits onlap as successive beds encroach over the erosion surface, whereas the top of the upper section exhibits offlap where successive beds have retreated seawards. Both sections exhibit downlap towards their seaward end as successive beds terminate against the erosion surface marking the base of the section. The downlap surface (maximum flooding surface) separating the two sections marks the change from onlap marking a transgressive environment, due to a rise in relative sea level, to offlap, marking a regressive environment, due to a fall in relative sea level. Note that a rise in relative sea level may either be caused by a global sea-level rise (i.e. a eustatic change) or by a tectonically induced depression of the basin floor. Parallel reflections indicate uniform rates of deposition, whereas a divergence of the reflections in a basin-wards direction indicates a differential subsidence of the basin floor.

Maritime and hinterland sequences

Vail *et al.* point out the importance of recognizing the distinction between maritime and hinterland sequences. A 'maritime sequence' consists of genetically related coastal and/or marine deposits in which the coastal element is controlled by the position of sea level whereas the deep-marine element is not. Thus cyclic changes exert an important control on the landward extent of maritime depositional sequences. A 'hinterland sequence', in contrast, consists entirely of non-marine deposits laid down on a site that is landward of the coast, where the deposition is only indirectly, if at all, influenced by the position of sea level. Such deposits appeared to be formed independently of the maritime sequences in general.

The authors propose that the depositional limits of onlap and toplap within the coastal facies of maritime sequences, as displayed in seismic sections, can be regarded as indicators of relative changes in sea level, as discussed below.

Sea-level changes

The third contribution in the series by the Vail group – by Vail, Mitchum and Thompson (1977b) – deals with the effects of sea-level changes and how they can be interpreted from seismic sections by the geometry of onlap of coastal deposits. The authors show how charts can be drawn up of the relative sea-level changes, which can then be used to produce a history of the fluctuations of sea level with respect to the land surface. They emphasize that such a relative sea-level change may be caused either by a eustatic rise in sea level, or by a fall in the land surface, or indeed by a combination of the two, and that consequently any such change in apparent sea level may be attributable to either a local, a regional or a global process.

Relative rise in sea level can be measured most accurately where littoral deposits (i.e. those deposited between low and high tides) can be detected on-lapping the depositional surface. However, in many cases these will have been removed by erosion. Where the relative rise is more rapid than the rate of deposition, transgression occurs, as shown in Figure 11.9A, as the shoreline advances landwards. This could result in the onlap of marine strata, in which case the extent of the sea-level rise could be more difficult to estimate. If the rate of sediment supply exceeds the amount required to fill the available space, regression occurs, and the shoreline retreats basin-wards, as shown in Figure 11.9B. If the two processes balance out, the shoreline will remain stationary, as in Figure 11.9C.

A relative fall of sea level is marked by a seawards shift of the coastal onlap, usually resulting in erosion. After a major fall the shelf would tend to be exposed to subaerial erosion, causing the sediments to be funnelled down channels cut through the shelf.

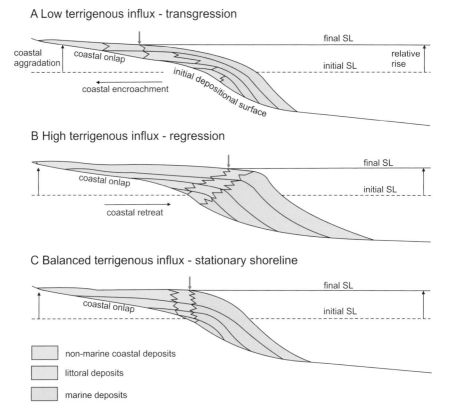

Figure 11.9 Transgression, regression and coastal onlap during relative rise of sea level. The rate of terrigenous influx determines whether (**A**) transgression or (**B**) regression occurs, or (**C**) the shoreline remains stationary. The red arrow denotes the shoreline position. After Vail *et al.*, 1997.

A 'stillstand' of sea level is where the sea level maintains an apparently constant position relative to the surface of deposition, and is indicated by coastal toplap, as shown in Figure 11.5A. In this case, each successive new sedimentary unit terminates in a progressively seawards direction.

Cycles of sea-level change

The authors define a 'cycle' of relative sea-level change as 'an interval of time during which a relative rise and fall of sea level takes place'. They note that such a cycle typically consists of a gradual rise, followed by a period of stillstand and then a rapid fall (Fig. 11.10). The abrupt fall in sea level at the end of a cycle would normally produce an unconformity, which would define the top of a depositional sequence, separating it from an overlying sequence. The greater the relative fall in sea level, the easier it is to recognize sequence boundaries by the onlap, downlap and truncation geometries that they display.

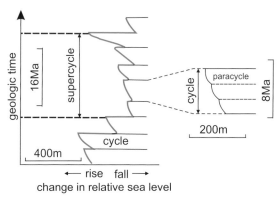

Figure 11.10 Chart illustrating cycles of relative sea-level rise. A cycle consists of a gradual rise in relative sea level to a stillstand, followed by a sudden fall, over a timescale of the order of c.8 Ma. Several cycles form a supercycle, over timescales of the order of c.16 Ma. Cycles may themselves contain several smaller-scale paracycles, measured over timescales of c.4 Ma. Modified from Vail *et al.*, 1997.

Each cycle may contain a single sequence. However, where there is a sufficient sediment supply, two or more separate sequences may be detected within a cycle. In such a case, when looked at in detail, the 'gradual' rise in the cycle may actually consist of a number of smaller-scale rapid rises and stillstands. These, termed 'paracycles', may not be recognizable seismically and may be confined to local areas of abnormally high deposition such as deltas, which may have no regional or global significance.

Construction of regional curves of relative sea-level change

Vail *et al.* conclude part 3 by explaining how sea-level curves may be compared with other regional data to produce a chrono-stratigraphic correlation chart. The process is explained in Figure 11.11, which is in three parts. Part A shows a typical seismic stratigraphic section consisting of five sequences, A–E. Part B is a chrono-stratigraphic representation of A, and part C is a chart showing the curve of relative rises, stillstands and falls represented in the stratigraphic section, constructed with respect to geologic time. In this chart, cycle A represents a highstand, when sea level is above the shelf edge, whereas cycle B represents a lowstand, when sea level is at its lowest position on the shelf during the deposition of a series of sequences. Cycles B–D are grouped into a super-cycle, defined by a gradual rise from the A–B boundary, interrupted by minor falls between B and C and between C and D, ending in a stillstand, followed by a major fall between D and E.

The highstands are the most likely times for trapping terrigenous clastic deposits in deltas on the shelf, whereas, during the lowstands, the clastics tend to bypass the shelf by being funnelled through submarine canyons onto the basin floor. The position of the shelf edge is shown as 0 metres on the chart (C), with highstands to the left and lowstands to the right.

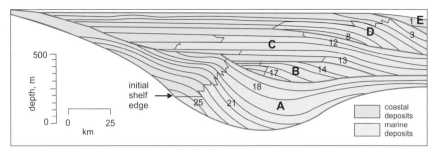

A Stratigraphic cross-section: ages in Ma

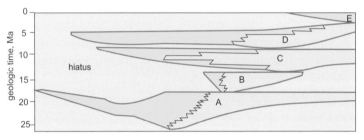

B Chrono-stratigraphic chart

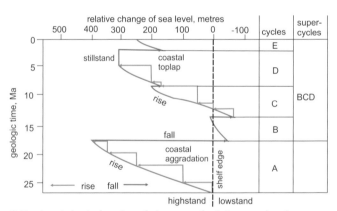

C Regional chart of cycles of changes of relative sea level

Figure 11.11 Procedure for constructing a regional chart of changes of relative sea level. **A** seismic section showing major reflectors, sequence boundaries and unconformities (in red); known dates in Ma. **B** chrono-stratigraphic chart of A. **C** chart of A drawn to show changes of relative sea level, in metres, against geologic time. After Vail *et al.*, 1977.

Global sea-level cycles

Part 4 of the series of contributions by Vail and colleagues (Vail, Mitchum and Thompson, 1977b) deals with cycles of relative sea-level change on a global

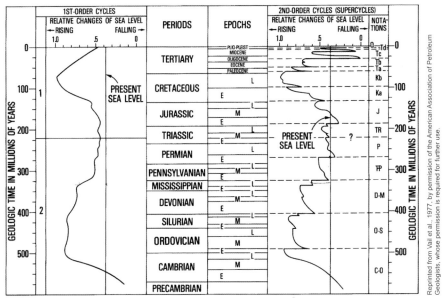

Reprinted from Vail et al., 1977, by permission of the American Association of Petroleum Geologists, whose permission is required for further use.

Figure 11.12 The Vail global sea-level chart, showing cycles of relative change of sea level for the Phanerozoic. From Vail *et al.*, 1977, with permission.

scale, and contains the much-quoted 'Vail sea-level curve'. Although global sea-level changes had been discussed previously by many geologists, including Suess and Stille in the early years of the twentieth century, the much expanded dataset available from the seismic records enabled a much more accurate sea-level chart to be constructed (Fig. 11.12).

The authors introduce the subject with the following statement:

'Cycles of relative sea-level change on a global scale are evident throughout Phanerozoic time based on the fact that many regional cycles on different continental margins are simultaneous and their relative magnitudes generally similar. Global cycles are records of geotectonic, glacial and other large-scale processes, they reflect major events of Phanerozoic history.'

They recognize three major orders of cycles superimposed on the global sea-level curve: first-order cycles with durations of 200–300 Ma, second-order with durations of 10–80 Ma, and third-order with durations of 1–10 Ma (not shown in Fig. 11.12). There are two first-order cycles recognizable in the Phanerozoic (Cambrian–Triassic, and Triassic–?), up to 14 second-order cycles and around 80 of the third-order cycles (excluding late Palaeozoic cyclothems).

The authors conclude that only geotectonic mechanisms are of sufficient duration and magnitude to account for the first-order cycles and for most of the second-order ones. It would appear that changes in volume or elevation of the mid-ocean ridge system, related to changes in the rate of sea-floor spreading, were the most likely mechanisms to explain these. However, the twin processes of glaciation and deglaciation were probably responsible for many of the third-order cycles and some of the second-order ones.

The Exxon team end their contribution with the following statement, which summarizes one of their most significant contributions:

> 'One of the greatest potential applications of the global cycle chart is its use as an instrument of geochronology. Global cycles are geochronological units defined by a single criterion – the global change in the relative position of sea level through time. ... As seen on the Phanerozoic chart [see Fig. 11.12], the boundaries of the global cycles in several cases do not match the standard epoch and period boundaries, but several of the standard boundaries have been placed arbitrarily and remain controversial. Using global cycles with their natural and significant boundaries, an international system of geochronology can be developed on a rational basis. If geologists combine their efforts to prepare more accurate charts of regional cycles, and use them to improve the global chart, it can become a more accurate and meaningful standard for Phanerozoic time.'

Later developments

The publication of AAPG Memoir 26 in 1977 led to the widespread adoption of the methods of seismic stratigraphy, not only in petroleum exploration but also in academic research in the fields of stratigraphy and sedimentology. Much of the subsequent work in these fields is summarized in the 1997 textbook by Andrew Miall. In the final chapter of his book, Miall lists a number of problems with the applicability of the worldwide sequence concept, which had attracted considerable controversy in the intervening years.

Terminology

The term 'sequence' as used by Sloss in 1963 refers to cycles of major sea-level change that are tens of Ma in duration. However, Vail and colleagues term these 'second-order cycles' and reserve the term 'sequence' for much smaller packages of strata representing sea-level cycles of only a few Ma in duration (i.e. 'third-order cycles'). The concept of 'allostratigraphy' was subsequently introduced to describe such sequences. The North American Commission for stratigraphic Nomenclature (NACSN) in 1983 defines the allostratigraphic unit as:

> 'a mappable stratiform body of sedimentary rock that is defined and identified on the basis of its bounding discontinuities.'

A hierarchy of such units was proposed, including 'allogroup', 'alloformation' and 'allomember' mimicking the respective lithostratigraphic units, but applied to sequences, each of which might contain several different lithological types, but which merely represent facies changes in a unified deposit or set of deposits.

Stratigraphic cycles and chronostratigraphy

The decision as to whether a particular sequence is regional or global is critical in determining whether an apparent sea-level change is eustatic, due to a worldwide change in sea level caused, for example, by the melting of polar ice caps, or is a merely local effect induced by tectonic movements. In practice, it is often difficult to correlate sequences owing to the lack of sufficiently precise dating. That being so, the Exxon system of first, second and third-order cycles is not recommended in view of the continuous range of frequencies now evident from the stratigraphic record.

Certain 'global' sequence boundaries with a periodicity of 200–500 Ma can be correlated with the assembly and break-up of supercontinents, and these can be assumed to relate to genuine eustatic sea-level changes. However, other major boundaries may be due to changes in climate and latitudinal position that are only continent-wide. Some may relate to changes in plate-kinematic patterns caused, for example, by rifting or collision that affect, and are correlatable between, adjacent plates but not globally.

Cycles with 10 Ma periodicity appear to be caused primarily by variations in ocean-basin volume generated by episodic sea-floor spreading rates, as suggested by the Exxon team. However, major sea-level changes may also have a thermal cause related to mantle convection: for example, a geoid bulge due to a mantle plume (see chapter 10). These may be continent-wide but would not be global.

Sequences with periodicities of *c.*1 Ma caused by repeated transgression/regression cycles are common in extensional basins. A special case of these are the Upper Palaeozoic 'cyclothems' of the northern hemisphere, which are attributed to sea-level changes caused by the repeated glacial and interglacial periods of the Gondwana glaciation in the southern hemisphere.

In practice, it is difficult to distinguish between the effects of the various processes that generate unconformities by means of their effects on sea level. Some of these processes are regional, others are global; some are rapid, while others are slower.

Miall concludes that sequence boundaries cannot be dated with sufficient accuracy to support the setting of precise ages to the 'global eustatic changes' that form the basis of the Exxon global cycle chart:

> 'The global accuracy and precision of the Vail curves is in question, and tectonic arguments are in as much danger as eustatic arguments of falling into the trap of false correlation and circular reasoning.'

Carbonate systems

The Exxon basic sequence model was introduced to describe a specific tectonic setting: namely the predominantly siliciclastic successions of the extensional continental margin. Therefore caution is required in extending it to other types of setting. The carbonate sedimentary systems differ from the siliciclastic in their response to sea-level changes, so that the Exxon system cannot be simply transferred to carbonates. For example, deep-water carbonates are more typical of highstand systems than lowstand ones, as would be the case for siliciclastic examples. Evaporite deposits likewise require major revision of the model.

Sediment supply

The model adopts an overly simplistic approach to the interaction between the three major controls on basin architecture: subsidence, sediment supply and sea-level change. The creation of accommodation space for sediments is governed more by tectonic uplift of the source area than by eustatic sea-level changes, as the Vail model implies. Moreover, sediment supply can also be affected by distant tectonic effects via major rivers that can cross entire continents.

Conclusions

Miall recommends acceptance of the broad principles of sequence stratigraphy while being extremely cautious in applying the global eustasy model:

> 'Peter Vail has started a legitimate revolution of great importance in stratigraphy, but we need now to separate the real advances and contributions from the ideas that have failed.'

As to its implications for petroleum geology, Miall concludes:

> 'Establishing the relationship of the sequence architecture to regional tectonic events is a critical component of basin analysis, whereas the global cycle chart is of no practical use in basin analysis and exploration...'

Postscript

Despite the limitations of the sequence stratigraphic model as originally set out by Laurence Sloss, Peter Vail and the Exxon geologists, it would not be an exaggeration to claim that it revolutionized the disciplines of stratigraphy and sedimentology. By the widespread use of seismic surveying, geologists were freed from the constraints of the outcrop, and were given a tool to enable reliable correlations to be made between outcrops and borehole sections – no longer tied to lithostratigraphic correlations that were known to be potentially, if not actually, diachronous. Using these techniques, exploration geologists have been able to vastly expand the known petroleum reserves by exploiting smaller deposits whose geometries have only been traceable through the use of seismic stratigraphy.

The last word should be left to Laurence Sloss himself, who in 1991 made the following statement:

> 'it is to be hoped that the cult of Neo-Neptunism, which demands that continents, their degree of freeboard [the amount by which a continent stands above sea level], and their cratonic basins and arches are subject to domination by eustasy, can be set aside as a rather quaint oversimplification.'

12

Gravity spreading

Historical background

'Gravity spreading' and 'orogen collapse' are relatively new concepts arising from the discovery that the later tectonic history of many mountain belts is dominated by extensional movements on faults and shear zones that have resulted in lateral spreading and thinning of the over-thickened crust of the orogens. A key paper on this topic was published by John Dewey in 1988, entitled *Extensional collapse of orogens*. However, the basic idea of extensional thinning or stretching of the crust is much older and built on knowledge of the fundamental role of the gravitational force in tectonic theory that has existed since the early days of geology. The term 'gravity spreading' appears in a 1987 paper (read at a Geological Society of London meeting in 1985) by L.H. Royden and B.C. Burchfiel:

> 'we propose gravitational spreading of the upper to middle crust of the High Himalayas within a primarily convergent regime as crustal material at mid-crustal levels is displaced southwards…we believe that the gravity spreading or gravitational collapse inferred for the Miocene Himalayas is a secondary effect superimposed on regional N–S compression.'

The authors acknowledge a similar idea developed in 1981 by B. Dahlmayrac and P. Molnar in relation to the Peruvian Andes, summarized by Royden and Burchfiel as follows:

> 'differences in confining pressure at constant depth (relative to sea level) tend to drive material from areas of higher topography to those of lower topography.'

It was the discovery of methods of establishing directly the sense of movement on faults and shear zones (see chapter 8) that enabled structural geologists to prove that many of the structures formerly thought to be thrusts (and which in many cases actually originated as thrusts) were extensional, and acted to thin, and reduce the height of, the orogenic edifice.

Early thoughts on the role of gravity

Despite the force of gravity being all-pervasive, well understood, and easily measurable, it received surprisingly little attention from structural geologists after the advent of plate tectonics until the 1980s. In contrast, the horizontal forces created by plate movements were regarded as the dominant influence on the generation of mountain belts and sedimentary basins. However, prior to the general acceptance of the plate-tectonic theory in the late 1960s, gravitational forces had figured much more prominently in the ideas of geotectonic theorists. Three geologists in particular, V.V. Beloussov, R.W. van Bemmelen and J.H.F. Umbgrove, who were prominent critics of continental drift, developed theoretical models in which vertical forces were the driving mechanism of tectogenesis.

The distinguished Dutch geologist Rein Van Bemmelen, famous for his three-volume treatise on the geology of Indonesia, attempted to explain the formation of mountain belts and geosynclines primarily as a consequence of vertical crustal movements in a series of papers from 1932 to 1952, culminating in his 1954 book *Mountain building*. For Van Bemmelen, the fundamental cause of orogenesis lay deep in the mantle, where density and volume changes produced vertical movements in the crust that were in turn acted on by gravitational forces to restore isostatic equilibrium. Beloussov held similar views: holding that geochemical processes related to the gradual differentiation of the material of the Earth's interior were the basic cause of crustal movements. Umbgrove took a different approach, believing that mountain building resulted from the collapse of geosynclines under the weight of their sedimentary and volcanic contents. The resulting influx of deep-crustal magma caused isostatic equilibrium to be restored with the rise of a mountain range. For all three of these geologists, and for others of the same period, the force of gravity was the fundamental driving mechanism behind Earth movements.

Gravity and tectonics

Isostasy

The concept of isostasy was introduced in the late nineteenth century, and is central to the understanding of how gravitational forces influence geological processes (see chapter 3 and Fig. 3.3). From measurements of the value of the force of gravity, g, around the Earth's surface, it has been shown that the Earth is in a state of approximate gravitational balance (i.e. isostasy), and thus the weight of any particular sector is similar to that of any other. Consequently, the less dense continents must attain a higher topographic level than the denser oceans if they are to have the same gravitational effect (i.e. the same weight). In other words, the continents are more 'buoyant' than the oceans. The process

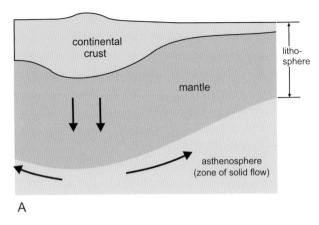

A

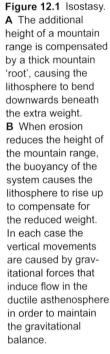

Figure 12.1 Isostasy.
A The additional height of a mountain range is compensated by a thick mountain 'root', causing the lithosphere to bend downwards beneath the extra weight.
B When erosion reduces the height of the mountain range, the buoyancy of the system causes the lithosphere to rise up to compensate for the reduced weight. In each case the vertical movements are caused by grav-itational forces that induce flow in the ductile asthenosphere in order to maintain the gravitational balance.

B

is made possible because weaker material in the upper mantle (i.e. within the asthenosphere) is able to flow in the solid state, enabling it to gradually adjust to the pressure of gravitational differences in the crust above.

Consequently, when a particular part of the crust is subjected to an additional load, for example by the creation of a mountain range, its base adjusts by sinking downwards and displacing mantle material sideways to form a 'mountain root' as shown in Figure 12.1. As in an iceberg, much more of the volume of the thickened crust of a mountain range lies beneath sea level than above. The reverse happens when crust is thinned to form an ocean basin: the top of the mantle bulges upwards and the surface is depressed, but the resulting basin is still in gravitational balance.

The effect of isostatic response to load is easily demonstrated by the evidence from the Scandinavian ice sheet, which existed during the most recent (Pleistocene) glaciation (see Fig. 3.4). During the glacial episodes, because so much water was contained in the ice sheets, which were up to 3 km thick in

Figure 12.2 Raised beach level backed by an old sea cliff, Lismore Island, western Scotland. IPR/73-34C British Geological Survey ©NERC. All rights reserved.

places, the mean sea level was much lower than it is now – by as much as 100 m. As the weight of ice was removed, the land slowly rose in response (a process known as 'glacial rebound') and the shorelines responded by rising above present sea level to form raised beaches (Fig. 12.2). These old beach levels can be recognized around the coasts of Scotland and Scandinavia, where they reach levels of up to 300 m above current sea level. By dating these, it has been possible to derive a value for the viscosity of the asthenospheric mantle of between 10^{21} and 10^{22} poise.

Gravitational gliding

Large sheets of rock, resting on a basal fault or other plane of weakness, if detached from their source, may experience sliding under gravity. This process is termed 'gravity sliding' or 'gravity gliding', and is a familiar result of earthquakes or rainstorms on unstable slopes. The down-slope progress of the sheet will be resisted by the rock mass at the end of the slope, unless the basal fault can escape onto the ground surface, in which case the sliding sheet may move as a coherent block, as in Figure 12.3A. If the basal fault does not reach the surface, the sheet may become internally deformed by folding and/or faulting because of the pressure exerted at its lower end, as in Figure 12.3B. Where the sliding nappe is unrestricted by such resistance, it may travel at high speed, comparable with that experienced in catastrophic landslides, but in the more usual case of restricted movement, the rate of deformation will correspond to that of 'normal' ductile flow in the solid state.

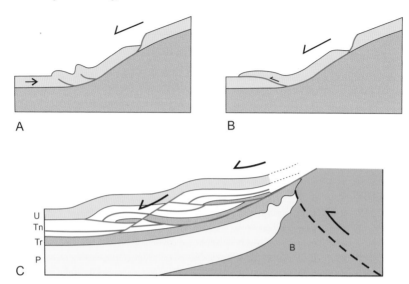

Figure 12.3 Gravity gliding. **A** The block sliding under gravity is resisted at the bottom of the slope causing deformation at the end of the moving block. **B** The downward progress of the moving block is aided by the basal fault cutting up to the surface. **C** R.H. Graham's interpretation of the Tineé nappes of the French Alps as gravity gliding structures. Many structures of this type have been re-interpreted as thrusts whose inclination has been reversed by subsequent uplift. The faults follow a detachment horizon within Triassic salt. B, basement, P, Pre-Triassic strata; Tr, Triassic; Tn, upper Jurassic; U, upper Cretaceous. After Graham, 1981.

Extending the gravity gliding model to a larger scale has proved controversial. Examples of gravity gliding nappes many kilometres in extent have been described from various mountain belts. The Western European Alps contain some of the best-known examples, one of which is illustrated in Figure 12.3C. The initiation of the gliding process in these situations is usually considered to be a response to crustal thickening due to compression in the central part of the mountain belt. It has been shown that gliding can take place on slopes of only a few degrees provided that the weight of the fault sheet is supported by a high enough fluid pressure within the fault zone. However, many examples of large-scale gravity gliding have been disproved on the grounds that the original slope of the slide surface was towards the interior of the orogen rather than towards the foreland, and that the nappes in question were actually thrust sheets whose direction of inclination had been reversed by subsequent uplift. To some extent the gravity gliding model has been superseded by the concept of gravity spreading (see below).

Hans Ramberg's contribution

Perhaps the most influential exponent of the role of gravity in tectonics in the pre-plate-tectonic era was Norwegian geologist Hans Ramberg. Like other believers in the fundamental importance of gravity, such as Beloussov and Van Bemmelen, he believed that the force of gravity was fundamental in tectonics, and he spent the greater part of his academic career conducting experiments that simulated actual geological situations. Unlike other researchers, however, Ramberg used materials whose physical qualities matched real rocks more closely because he conducted his experiments with a centrifuge, allowing the force of gravity to be scaled up by a factor of $c.2000$.

Hans Ramberg (1917–1998) was born in Norway and obtained his PhD in Oslo in 1946. However, he spent the first half of his academic career in Chicago, from 1948 to 1961, including three years at the Carnegie Institution, and the second half at Uppsala University in Sweden, from 1961 to 1982, where he founded the tectonic laboratory that currently bears his name: the Hans Ramberg Tectonic Laboratory. He was succeeded there in 1983 by Chris Talbot, himself well known for his studies on salt tectonics.

Ramberg's experiments

Ramberg's first experiments were carried out using a large-capacity centrifuge at the Enrico Fermi Institute in Chicago and published in 1963. The advantage gained by scaling up the gravitational force by a factor of over 2000 meant that rock structures, which in nature would take thousands or millions of years to develop, could be simulated in minutes. The first series of experiments was designed to simulate natural occurrences of domes or pipe-like structures and associated basins or synclines formed in the solid state. Various types of wax, modelling clay and putty were used, but the most realistic results were obtained using domes of 'bouncing putty' with a layered overburden consisting of alternate layers of bouncing putty and modelling clay. This produced realistic scale models of salt dome evolution (of which natural examples had been well studied) over times of the order of 10^3 seconds, which scale up to 8.8 Ma in geological time. The materials used for the source layer had an effective viscosity of 3×10^7 poise and the overburden of 7×10^5 poise. These equate respectively to 10^{18} and 10^{16} poise when scaled up, and are reasonable approximations to the calculated viscosities of natural rocks within the crust. Figure 12.4 illustrates the results of some of these experiments.

Salt tectonics

Salt has a low viscosity compared to most rocks and is therefore able to flow under relatively small stress differences and at geologically fast rates. Common salt (halite) and anhydrite are the commonest types of evaporite, formed by evaporation in warm shallow seas. Evaporite layers form convenient glide planes for thrust sheets in the Alps, and in northern Germany a Permian salt layer has

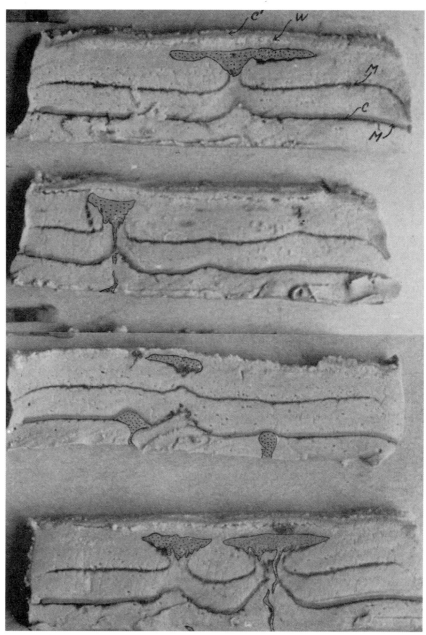

Figure 12.4 Ramberg's centrifuge model showing the rise of a layer of less dense, buoyant material through an overburden of denser material, simulating the behaviour of salt domes. From Ramberg, 1963, with permission.

developed into numerous salt domes and 'diapirs'. The nature of these structures and the way they develop is illustrated diagrammatically in Figure 12.5. The low-density salt layer is overlain by higher-density sedimentary layers that

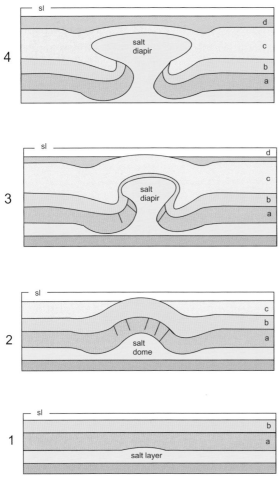

Figure 12.5 Evolution of a salt *diapir*. The salt layer is gravitationally unstable since it is less dense and has a lower viscosity than the overlying rock. Stage (1): a slight thickening of the salt layer is present at the site of the future diapir; (2) the gravitational load of the overlying layers *a* and *b* squeeze the salt sideways into a dome shape; extensional faults develop in the overlying competent layer *a* while deposition of layer *c* onlaps the sides of the dome; (3) with further doming, the salt breaks through the faulted layer *a* and spreads sideways into layer *b* to become a diapir; layers a, b and c are domed upwards and distorted into two asymmetrical synclinal structures at the sides of the diapir, while deposition of layer *d* takes place around its sides; (4) the diapir is now fully developed by spreading sideways into layer *c*, where it will now be more stable gravitationally. Based on Trusheim (1960).

exert a downward gravitational pressure on the salt. A slight local thickening of the salt layer can focus the salt into a dome-like or anticlinal structure, since the salt can be squeezed sideways by the gravitational load on either side (Figure 12.5.1). Nearly a kilometre of sediments is required to produce a sufficiently large gravitational pressure to start the process, but having started, the salt will tend to proceed upwards due to its buoyancy, aided perhaps by extensional fracturing of overlying layers. Once the salt has breached the overlying layer, it forms a salt diapir (Figure 12.5.3). When it reaches a level where it is more gravitationally stable, it will spread sideways to form a mushroom-like structure, as shown in Figure 12.5.4, which may eventually become completely detached from its roots. The diagrams of Figure 12.5 are based on natural examples from northern Germany published by Trusheim in 1960, and are remarkably similar to the results of Ramberg's models.

Ramberg used his experimental results to suggest that natural geological structures such as granite domes, ocean ridges and the arc–trench system of the Pacific rim had been produced solely by gravitational forces acting on instabilities within the crust and upper mantle produced by density contrasts. Thus, for example, a layer of salt beneath an overburden of sandstones or shales has an inherent buoyancy that only requires a small initial bulge or discontinuity to cause it to move upwards towards the surface. Similarly, density contrasts caused by warmer mantle material could provide the motive force for the rise of ocean ridges.

Application to orogenic belts

A later set of model experiments was undertaken by Ramberg to simulate the structures of orogenic belts such as the Alps and the Scandinavian Caledonides, and are illustrated in his 1967 book. He noted that there were many factors that caused gravitational instabilities, including the emplacement of dense basic volcanic material above much lighter sedimentary layers within geosynclines. By designing models with similar contrasts in density, he was able to reproduce structures that are superficially similar to those observed in orogenic belts (Fig. 12.6).

While Ramberg acknowledged that gravity-induced rearrangement of an unstable mass distribution could also be achieved as a result of lateral compression, his experiments were designed to show that such a rearrangement

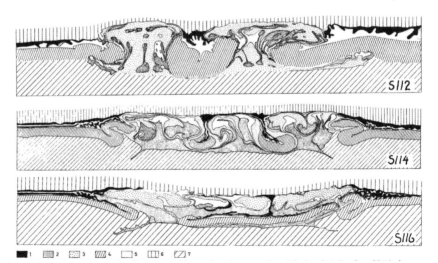

Figure 12.6 Ramberg's centrifuge models simulating orogenic structures. The less dense layers 2, 3 and 5 (ρ = 1.35, 1.25 and 1.14 respectively) have risen buoyantly through the denser layer 4 (ρ = 1.87) to form diapiric structures. Note that layers 2, 3 and 5 are confined to the central 'orogenic' part of the model and are absent on the 'forelands'. From Ramberg, 1967, with permission.

was possible through the operation of gravitational and buoyancy forces alone without the application of any externally applied compressive force. The concept of gravity spreading is clearly implicit in this work.

Crustal extension

The Basin and Range Province

The important role of gravity in crustal extension became obvious from the studies of the Basin and Range Province, which lies within the North American Cordilleran orogenic belt. At its broadest, between the Sierra Nevada Mountains in the west and the Sevier ranges in the east, the province is about 800 km wide, and is characterized by a series of narrow mountain ranges separated by wide flat valleys. The central and northern parts of the province are generally over 2000 m in altitude, with the individual mountain ranges rising to over 3000 m. Further south, the land is much lower. The valleys are mostly hot dry deserts, including the notorious Death Valley near the western margin of the province.

Although the crust is of normal thickness, the province experiences a higher than normal heat flow, and the mantle lithosphere is abnormally thin. This explains the relatively high altitude and structural weakness of the Basin and Range, which contrasts with the strong cool lithosphere of the Colorado Plateau to the east. The province became a distinct tectonic entity during the Jurassic when lateral compression created an over-thickened crust throughout the whole Rocky Mountain belt, including both the Basin and Range and the Colorado Plateau. After the Mid-Eocene, the subduction regime changed: a volcanic arc was formed along the western margin of the province, and resulted in the initiation of a back-arc extensional regime on the upper plate, but which excluded the Colorado Plateau. This extensional phase lasted until around 25 Ma ago, when the subduction zone was replaced by the San Andreas Transform Fault.

Wernicke's extensional model

An important paper published by Brian Wernicke in 1981 provided the key to understanding how large extensions could be accommodated by faulting, and the Basin and Range Province is now regarded as the classic example of extensional tectonics. The mountain ranges correspond to uplifted blocks (horsts) bounded by normal faults, and the valleys to the intervening structural depressions (graben). However, only a limited amount of extension can be achieved by a set of normal faults, and Wernicke showed that for further extension to take place, the faults need to be rotated into a more gently inclined attitude, as seen in Figure 12.7. This has the result of also rotating the fault-bounded blocks, so that originally horizontal strata within the blocks also become inclined. The

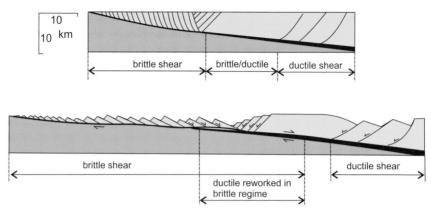

Figure 12.7 Wernicke's model for large extensions. Steep normal faults rooting on a basal detachment accommodate extension by rotating into a shallower inclination forming a series of fault blocks like a set of dominoes. Those at upper crustal levels form under brittle conditions but at deeper levels the basal detachment becomes a ductile shear zone and the steep faults bend towards it. As the extension proceeds, sections of the detachment that were formerly within the ductile sector rise up to higher levels as the crust thins and the ductile fabrics are overprinted by brittle ones. After Wernicke, 1981.

individual fault blocks rotate above a basal fault or shear zone that acts as a décollement horizon. Some of the faults may be 'listric': that is, they curve into a shallower inclination at depth to enable the blocks above them to rotate and thus allow larger extensional displacements to take place. Many of the uplifted blocks expose the Precambrian basement and are known as 'metamorphic core complexes'; such complexes are now known to be typical of extended regions and illustrate the influence of buoyancy forces in an extensional regime.

Gravitational effects of large extensions

Extensional displacements on large normal faults are gravitationally unstable (Fig. 12.8). The effect can be simulated by floating a wooden block in a tank of water. When the block is faulted, the originally horizontal surface must tilt to retain equilibrium, as shown in Figure 12.8-3. In natural examples, the gravitational response of the extensional crustal thinning is to produce an antiformal fold in the hangingwall (often termed a 'rollover') and a synformal fold in the footwall (12.8-4). In this way, the gravitational imbalance that would have resulted if no bending had taken place is smoothed out.

Large extensional displacements of the crust result in local thinning, which in turn reduces the weight of that particular section of the crust. To maintain gravitational equilibrium, it is necessary for the underlying mantle material to rise isostatically so that the total weight of that section remains the same. Thus in Figure 12.9A, the mantle material beneath the detachment fault has bulged upwards to balance the effect of the thinned crustal material. Figure 12.9B–C

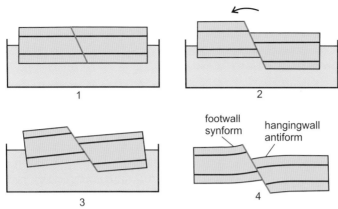

Figure 12.8 The gravitational effect of a large normal fault can be visualized by considering the effect of a wooden block floating in a tank of water (1); once faulted, the block is unstable and must tilt (3) to maintain equilibrium; (4) the tilted sections of the block are represented in nature by a hangingwall anticline (*rollover*) and a footwall syncline.

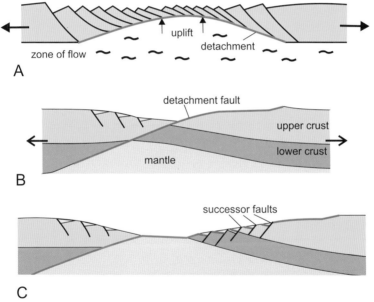

Figure 12.9 A Cross-section through the crust showing how large amounts of extension and thinning of the upper crust coupled with solid-state flow in the lower crust can lead to arching up of the lower-crustal material. **B, C** Cartoon sections (after Reston, 2007) showing how the crust may be stretched and thinned to form an oceanic basin by the development of a low-angle normal fault that acts as a detachment horizon; B, as the hangingwall moves to the left, the footwall rises to retain gravitational equilibrium; C, the mantle, roofed by the detachment, is now at the surface; successor normal faults cut through the footwall into the lower crust, which is now brittle.

shows how stretching of the crust by means of a low-angle normal fault may ultimately lead to the rise of the mantle to the surface and the creation of an ocean basin. As the structure evolves, the original detachment fault becomes inoperative in the footwall and further extension takes place along steeper 'successor' normal faults.

Extension in the Tibetan Plateau

The vast Tibetan Plateau extends for over 2400 km from west to east and is 1300 km across at its widest. It is everywhere over 4000 m in height and includes several impressive mountain ranges containing peaks over 7000 m high. The southern margin of the plateau is defined by the Himalayan and Karakoram Ranges, while the northern boundary is the Altyn Tagh Fault, which separates the plateau from the Tarim Basin (Fig. 12.10). The north-eastern boundary is defined by the Nan Shan fold belt at the southern edge of the Gobi Desert. This part of the Eurasian plate has been less obviously affected by the Himalayan compressional deformation, part of which is concentrated along the suture zones between the several separate terranes that had previously accreted to Central Asia.

The Himalayan deformation has also been accommodated by movements along a network of strike-slip faults, which form a conjugate set: a sinistral set

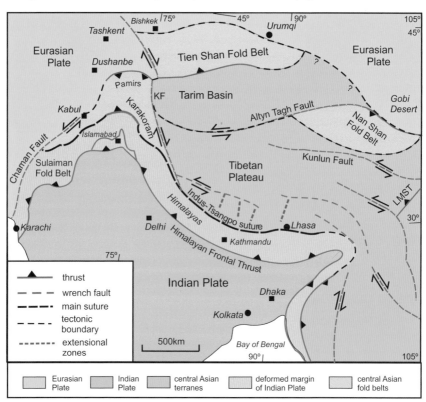

Figure 12.10 Regional tectonic setting of the Tibetan Plateau. KF, Karakoram Fault; LMST, Longmen Shan thrust belt.

varying in orientation from WNW–ESE to ENE–WSW, and a dextral set varying from N–S to NNW–SSE. Towards the east, the faults of both sets curve round into a more N–S orientation as the belt of deformation turns southwards into Burma and Indo-China. Numerous north–south oriented graben systems also indicate significant E–W extension and this, coupled with the movements on the numerous minor conjugate wrench faults, has been interpreted to indicate that much of the north–south convergence between India and Asia has been accommodated by the sideways extrusion of Asian crust.

GPS measurements have enabled accurate movement vectors across the orogen to be calculated (Fig. 12.11); these range generally between 5 and 15 mm/a, confirm the lateral extrusion model, and also indicate a gradual diminution of flow velocity northwards; there is thus no clearly defined northern margin to the deformation resulting from the collision as there is in the south. Slip rate measurements along the wrench faults are also typically in the range 5–15 mm/a.

This pattern of distributed deformation, achieved mainly by faulting, applies only to the upper brittle crust. Beneath this, the middle crust, being warmer and more ductile (see below), is believed to have deformed in a more continuous manner, employing shear zones rather than discrete faults.

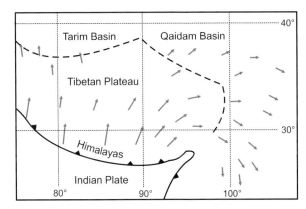

Figure 12.11 Velocity vectors (red arrows) showing direction and relative movement of the deformed part of the Eurasian Plate relative to stable Eurasia. Note the flow pattern in the eastern sector indicating the eastern 'escape' of Tibetan crust. After Searle *et al.*, 2011.

The origin of the plateau

There has been a vigorous debate that has lasted for many decades about the reason for the high elevation of the plateau. Orthodox plate-tectonic theory originally suggested that the explanation was the under-thrusting of Asia by the Indian Plate, causing a doubling up of the Tibetan crust. However, this would imply a very shallow-dipping subducting slab extending for over 1000 km, which mechanically seemed unlikely. Seismic data show that the present base of the Indian crust descends from about 40 km depth at the orogenic front to 70 km beneath the Indus–Tsangpo Suture and remains around that level for

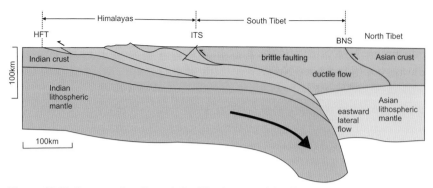

Figure 12.12 Cross-section through the Himalayas and the Tibetan Plateau, showing the contrasting nature of the crust and upper mantle between the Indian and Asian Plates. Note that the Asian crust is much thicker than the Indian, and that part of the Indian lithosphere beneath Tibet is apparently 'missing'. BNS, Bangong–Nujiang Suture; HFT, Himalayan Frontal Thrust; ITS, Indus–Tsangpo Suture. After Searle *et al.*, 2006.

about 200 km across South Tibet before it disappears to be replaced by the quite different structure of the Tibetan Plateau (Fig. 12.12). As the subducting slab has now been shown to extend for only a relatively short distance beneath the southern part of the Plateau, as indicated in Figure 12.12, it cannot be responsible for the plateau uplift.

Various indirect geophysical techniques have revealed that the crust of the Tibetan Plateau is considerably warmer, and therefore much weaker, than the Indian equivalent, and that its thickness is more than double, varying from 70 km in the south to around 60 km in the north. Moreover, as a result of the different physical properties in the lower part of the crust, the brittle, fault-dominated tectonics of the upper crust would be expected to be replaced in the lower crust by a pattern of general eastwards flow.

The behaviour of the lithospheric mantle beneath Tibet is more difficult to determine. Because of the marked increase in strength between the lower crustal and uppermost mantle material, the latter would not be expected to behave in the same ductile fashion. Nevertheless the seismic properties of the Tibetan mantle contrast with those of the Indian lithospheric mantle and are consistent with a model of mantle flow similar to that inferred for the crust, at least for the material below the topmost part. The 'missing' Tibetan mantle lithosphere caused by the under-thrusting of the Indian lithosphere, evident from Figure 12.12, could thus be explained by the absorption of the lower part into the asthenosphere, leaving the total lithospheric thickness more or less unchanged.

The conclusion drawn from these observations is that much of the compressive stress generated by the Indian collision has been accommodated by shortening of both the crust and lithospheric mantle of the Tibetan Plateau, accompanied partly by crustal thickening and partly by sideways transfer of material, mostly to the east and southeast. The height of the plateau can be explained by its warmth and thickness, and its relative flatness by the fact that general isostatic equilibrium can be achieved by the ease of flow in the lower crust. The high heat flow is attributed to large quantities of igneous material within the crust, mostly arising from the pre-collision subduction process.

The Indian lithosphere, in contrast, is believed from seismic evidence to be relatively cool and strong, providing a broad semi-rigid slab that has underthrust the Asian plate for a distance of over 200 km north of the Indus–Tsangpo suture. The northward movement of this slab is the main driver for the deformation of the orogen. The current convergence rate between the Indian and Eurasian plates is about 50 mm/a. Of this, approximately half is taken up by shortening across the Himalayan fold-thrust belt and half by the shortening and eastwards extrusion of the Tibetan lithosphere.

Subduction rollback

This mechanism, also known as 'trench roll-back' or 'slab roll-back', was discussed in detail in chapter 9 (see Fig. 9.6). It is a process in which the upper plate of a subduction zone is extended and thinned due to the gravitational pull of the subducting slab. The downward force exerted by the descending dense oceanic lithosphere also acts to move the position of the trench backwards along the subducting plate, provided the inclination of the slab is more than 45°. Subduction rollback is a major cause of crustal extension, and its effects can be demonstrated by the development of numerous examples of back-arc basins around the modern circum-Pacific orogenic belt system.

Gravitational spreading

The basic idea behind gravitational spreading is that the orogen-building process, involving perhaps many thrust sheets piled on top of one another as a result of lateral compression, is gravitationally unstable. Although the thickened crust may be in isostatic equilibrium, gravitational forces will act to reduce the thickness to that of the surrounding crust by producing outward-directed stresses. This important principle was explained in a 1981 paper by Martin Bott and Nick Kusznir. Thus, during orogenesis, as the crust is compressed laterally and extended vertically, gravitational forces will eventually respond to reduce crustal thickness by the lateral spreading of the over-thickened crust (Fig. 12.13).

Whereas gravity gliding relies on a concentration of movement on a basal fault, gravity spreading implies largely ductile deformation throughout the moving sheet in response to a gravitationally derived stress state. On the scale

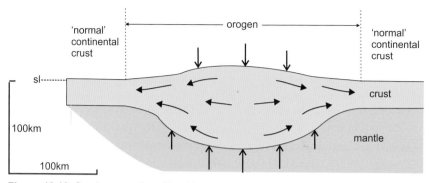

Figure 12.13 Gravity spreading. The thickened crust of an orogenic belt is gravitationally unstable. The resulting stresses squeeze the over-thickened crust, which flows laterally outwards.

of the whole mountain belt, the latter process has also been described as 'orogen collapse' (see below). On this scale, the over-thickened crust can be viewed as if it is a ductile medium responding to the gravitational stress, which acts to squeeze the crust, which 'flows' sideways in order to restore the more stable gravitational state of 'normal' crustal thickness. In detail, of course, the resulting deformation will be highly variable: brittle fault-controlled in the upper regions and more ductile creep in the lower.

One of the advantages of the lateral spreading model is that it provides a mechanism for generating outward (and upward) directed thrust sheets without necessarily relying on the compressional stress produced by convergence. Thus gravitationally generated thrusting could continue after active convergence between opposing plates had ceased.

Mantle thinning

One of the problems in understanding the Tibetan Plateau was how the lithospheric mantle beneath the thickened crust reacted to the convergence of the opposing plates. In a 1981 paper, Houseman, McKenzie and Molnar proposed that during crustal shortening, the underlying lithospheric mantle must also shorten and consequently thicken, causing relatively cold, dense lithospheric material to descend into the warmer, less dense convecting asthenosphere. This material, which the authors term the 'thickened boundary layer', may become unstable, detach, and sink into the asthenosphere, to be replaced by warmer asthenospheric material. The authors modelled this instability for various physical boundary conditions, and concluded that the boundary layer could be removed in times consistent with or shorter than the duration of the convergence (30–50 Ma). The consequence of this would be that the overlying crust and uppermost mantle would warm up rapidly, contributing to regional metamorphism and the generation of late-orogenic igneous activity.

John Dewey on orogen collapse

Dewey's 1987 synthesis, *Extensional collapse of orogens*, integrated existing models of gravitational spreading, subduction roll-back and lithospheric mantle thinning to explain the occurrence within the central parts of many orogenic belts of extensional regions that occur either as high plateaux or as low basins. According to him, extension of the continental lithosphere is driven by two horizontal tensional forces, one caused by isostatically compensated uplifts, as proposed by Bott and Kusznir, and the other by subduction roll-back.

Several well-known examples from the Mesozoic–Cenozoic orogenic belt system together with a few from older orogenic belts were chosen as illustrations, focusing especially on the Tibetan Plateau and the Basin and Range Province (see above) as examples of gravity spreading, the Aegean Sea as an example of slab roll-back, and the Alboran basin within the Betic–Rif belt of mantle thinning (Fig. 12.14).

Dewey proposed a five-phase evolutionary tectonic history for mountain belts.

- Phase 1: plate convergence causes lithospheric and crustal shortening and thickening, with a decrease in geothermal gradient and an absence of magmatism. This stage leads to a crustal thickness of 65 km and an elevation of 3 km. The elevation is isostatically compensated.
- Phase 2: thermal re-equilibration causes a slow thinning of the lithospheric mantle and a consequential increase in geothermal gradient resulting in prograde metamorphism and minor igneous activity.
- Phase 3: 'catastrophic' erosion of the lithospheric mantle results in rapid uplift, a rapid increase in geothermal gradient, high-temperature metamorphism, emplacement of granites and the initiation of crustal extension.
- Phase 4: a further increase in geothermal gradient causes lithospheric extension and thinning leading eventually to orogen collapse. Where the boundary forces are compressional, extension is balanced by thrusting, but where extension is dominant, subsidence may continue below sea level to create marine basins.
- Phase 5: post-extensional decrease in geothermal gradient causes both the lithospheric mantle and the crust to return to their 'normal' thickness by isostatic recovery.

Dewey's observations on Tibet

The Tibetan Plateau is taken by Dewey as the type example of the first stages of the extensional collapse process (i.e. phases 1–3). He remarks that as the grav-itational force increases with increasing crustal thickness and elevation, the

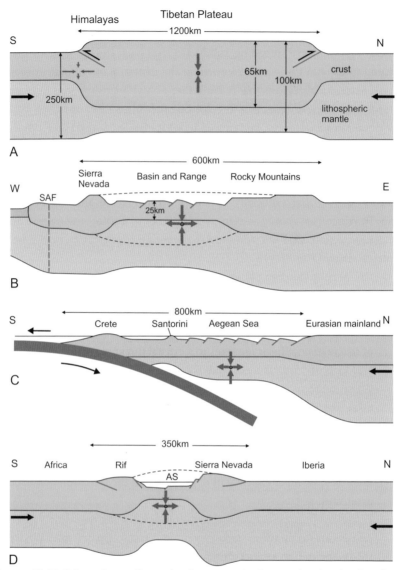

Figure 12.14 Schematic sections showing several stages of extensional collapse. **A** Himalayas and Tibetan Plateau. **B** Sierra Nevada–Basin and Range to Rocky Mountains. SAF, San Andreas Fault. **C** Aegean Sea. **D** Rif to Betic Alps across the Alboran Basin. AS, Alboran Sea. The black arrows represent movement directions; the green represent stress conditions (note that in A, the extension direction is eastwards, out of the page); red lines are faults; black dashed lines represent the position of the pre-extension crust. After Dewey, 1988.

strength of crustal material restricts the likely elevation of mountain belts to about 3 km above base level, after which they will be subject to gravity spreading. The ratio between the size of the force required to drive the convergence and the gravitational force is therefore critical in determining how the orogen interior responds. He notes that as a result of the collision of India with Asia, the thinner and weaker Asian lithosphere has doubled in thickness beneath the Tibetan Plateau, causing the plateau to be uplifted by about 2 km to reach its present elevation of *c*.5 km (Figure 12.14A).

As noted above, the plateau was subjected to E–W extension by means of a combination of strike-slip faulting and N–S graben formation. Outward-directed thrusting, however, was confined to the perimeter of the plateau – in the Himalayas, the Pamirs, the Nan Shan and the Longmen Shan ranges – and mainly affects elevations less than 3 km in height. This suggests that while the higher elevations of the plateau were experiencing extension, the lower parts were thickening and seemed to be collapsing radially. He concludes that because the extension of the plateau is restricted to an E–W orientation, and material is allowed to escape eastwards, plate convergence rather than orogenic spreading must be the dominant factor here. The high elevation and relative warmth of the plateau is attributed to the thinning of the shortened lithospheric mantle as suggested by the model of Houseman, McKenzie and Molnar.

The Peruvian–Bolivian Andes are regarded as a similar case. Here, the leading edge of the South America plate is being driven westwards by the expansion of the South Atlantic. A thickened crust has experienced recent rapid uplift and extension at elevations of over 3 km whereas compressional structures in the form of thrust belts only occur in regions of lower elevation.

The Basin and Range Province

Dewey regards the Basin and Range Province, described above, as illustrating phase 4 of his extensional collapse process, in which the collapse follows closely upon the shortening phase. The convective thinning of the lithospheric mantle root is suggested as a plausible mechanism for both the uplift and the extensional collapse (Fig. 12.14B).

The Aegean Basin

The Aegean Sea occupies a large extensional basin situated between the Hellenide chains to the northwest and the Tauride and Pontide belts of Turkey (Figs 12.14C, 12.15). The basin is bounded to the south by the Hellenic volcanic arc, which is seismically active. The site of the subduction zone is concealed beneath the Mediterranean Ridge, which occupies a broad zone in the central part of the eastern Mediterranean Sea and is interpreted as an accretionary prism that has accumulated on the site of the subduction zone over a long period of time, filling in the original ocean trench.

The southern part of the Aegean Sea, north of the volcanic arc, represents a back-arc basin, and the study of the active extensional faults, both in the Aegean islands and on the southern Greek mainland, indicates that the Hellenic Arc as

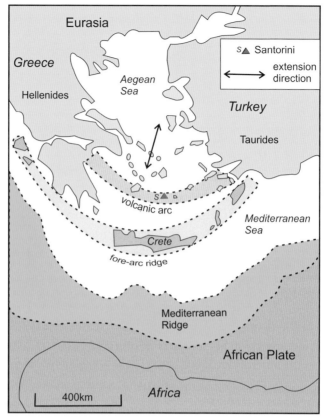

Figure 12.15 The Aegean Basin. The Aegean Sea is an active extensional basin situated between the Hellenides of Greece and the Anatolide–Tauride belt of Turkey. The extension is attributed to the southward roll-back of the subduction zone marking the northern margin of the African Plate and is concealed beneath the Mediterranean Ridge. After Okay, 2000.

a whole has experienced extension both in an approximately NNE–SSW direction, perpendicular to the arc, and also parallel to the arc as it has expanded outwards and stretched. The outward migration of the subduction zone, which has occurred over a period of at least 50 Ma, from the Eocene Epoch to the present, has resulted in crustal thinning leading to the depression of the whole Aegean Sea Basin.

This process, which is common to many present-day island arc systems, is attributed to the gradual outwards retreat of the subduction zone due to the process of trench, or slab, roll-back, illustrated in Figure 12.14C. This basin, along with several other similar basins within the Mediterranean sector of the Alpine-Himalayan orogenic belt, is regarded by Dewey as a type of orogen collapse where, in phase 4 of his tectonic sequence, the extensional boundary force generated by the roll-back has overcome or superseded the effects of plate convergence.

The Betic-Rif belt

This sector of the Alpine–Himalayan orogenic belt occupies the extreme western end of the Mediterranean, where the Betic mountain belt of southern Spain swings round in a 150° arc across the Straits of Gibraltar to join the Rif belt in northern Morocco (Fig. 12.16). The two arms of the arc enclose a marine basin occupied by the Alboran Sea. The problem pointed out by Dewey in interpreting the existence of this basin is that the convergence between Africa and Iberia is insufficient to account for all the outward-directed thrusting in the Betic and Rif belts. The present crustal thickness in the Alboran Basin is only about 15 km, indicating a large amount of extensional thinning. The simple solution of gravitational collapse as in the Tibetan model cannot be applied because the extensional basin is lower than the thrust belts and thus cannot cause them. Moreover, the extensional grabens have an E–W trend, which cannot explain the westwards thrusting of the Gibraltar Arc.

Dewey concludes that the base of the lithospheric mantle root originally thickened during the compressive phase, has been resorbed into the asthenosphere, and that the resulting isostatic response to the crustal thinning has depressed the surface of the crust, as shown in Figure 12.14D. However, it has subsequently been proposed (by Zeck in 1999) that the E–W extension has been caused by slab roll-back due to the eastwards retreat of the subduction zone that forms the western margin of the oceanic part of the African Plate (now located off Sicily).

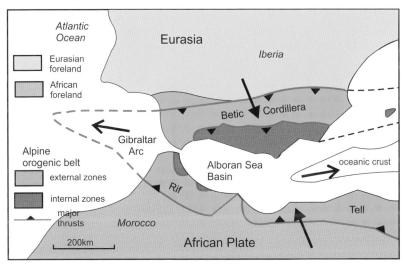

Figure 12.16 The Alboran Basin. Convergence between the Eurasian and African Plates has resulted in the Betic, Rif and Tell orogenic belts, the central part of which has collapsed to form the Alboran Basin. The Alboran crust is escaping westwards to form the Gibraltar arc and eastwards due to roll-back of the subduction zone (now located in eastern Italy and Sicily). Based on Sans de Galdeano, 2000.

Postscript

The significance of Dewey's paper is that it sets the gravity spreading model into a general theory of orogenesis, which includes an explanation of how the structural, metamorphic, and igneous history of orogenic belts can be accounted for by the changing balance between the compressional forces of plate convergence and the extensional forces generated by gravity. However, the gravity spreading model builds on concepts of the role of gravity in tectonics developed by earlier workers such as Hans Ramberg, Walter Elsasser and Martin Bott, and also incorporates insights on the geothermal role of the mantle in crustal behaviour by Dan McKenzie (e.g. McKenzie, 1978).

A more recent interesting development of the gravity spreading model is the concept of 'channel flow', to which a complete volume in the Geological Society of London Special Publications series was devoted (Law *et al.*, 2006). The channel flow idea, originally applied to the Himalayas, is that a warmer, more ductile layer of the crust is squeezed outwards and upwards towards the foreland beneath the gravitational load of cooler crustal material above. In the case of the Himalayas, the ductile channel consists of the metamorphic core known as the Greater Himalayan sequence, which is bounded above and below by shear zones with opposing shear sense.

Appendix

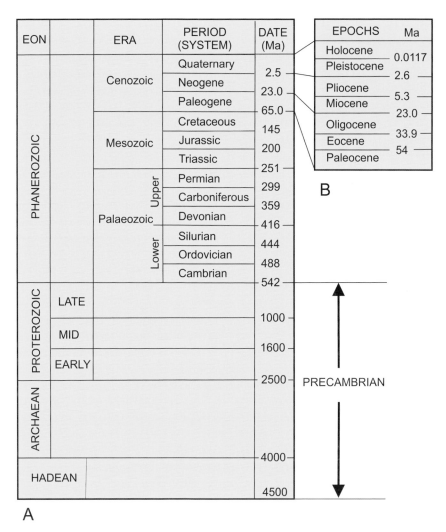

EON		ERA	PERIOD (SYSTEM)	DATE (Ma)
PHANEROZOIC		Cenozoic	Quaternary	2.5
			Neogene	23.0
			Paleogene	65.0
		Mesozoic	Cretaceous	145
			Jurassic	200
			Triassic	251
	Upper	Palaeozoic	Permian	299
			Carboniferous	359
			Devonian	416
	Lower		Silurian	444
			Ordovician	488
			Cambrian	542
PROTEROZOIC	LATE			1000
	MID			1600
	EARLY			2500
ARCHAEAN				4000
HADEAN				4500

PRECAMBRIAN

A

EPOCHS	Ma
Holocene	0.0117
Pleistocene	2.6
Pliocene	5.3
Miocene	23.0
Oligocene	33.9
Eocene	54
Paleocene	

B

Main sub-divisions of geological time. **A** Geological time units – eons, eras and periods; not to scale. **B** Epochs of the Cenozoic Period.

Glossary

A

accretionary prism (or complex): an accumulation of **clastic** sediments and volcanic debris occupying the trench and **continental slope** at an **active continental margin**, and typically deformed as a result of the subduction process.

acid (igneous rock): containing quartz, formed from a magma that is oversaturated in silica.

active continental margin: corresponding to a **subduction zone** or **transform fault**.

albite: sodium-rich **plagioclase**.

Algonkian: obsolete term used for early Precambrian basement rocks in North America.

allostratigraphy: term used to describe the stratigraphy of sedimentary sequences that are relatively small in scale (e.g. lasting for only a few Ma), and which may be contained within larger-scale, regionally recognizable sequences.

Alpine orogeny: series of orogenic episodes affecting southern Europe, resulting from the collision of Eurasia and Africa during the Cenozoic Era.

amphibole: member of a group of ferromagnesian silicate minerals with complex compositions; hornblende is the commonest variety.

amphibolite facies: metamorphic assemblage characterized by high temperatures and moderate pressures.

andesite: fine-grained **intermediate igneous** rock.

aplite: fine- to medium-grained acid igneous rock, occurring as narrow veins within a coarser-grained, typically granitic igneous or metamorphic host.

aseismic: lacking in earthquake activity.

asthenosphere: weak layer beneath the **lithosphere**, capable of solid-state flow, over which the **plates** move (see Fig. 6.10).

attenuation (of **seismic waves**): progressive diminution of wave amplitude with distance from the source.

augen: lens-shaped body (e.g. a deformed large crystal) within a **schist** or **gneiss**.

B

back-arc (basin): (sedimentary basin) situated on the upper plate of a **subduction zone** behind a volcanic arc.

back-arc spreading: process of forming oceanic crust behind a volcanic arc by extension of the upper plate above a **subduction zone** (see Fig. 9.6).

balanced section: reconstructed structural cross-section in which each measured bed length is equal.

basalt: fine-grained **basic** igneous rock.

baselap: **lapout** at the lower boundary of a depositional sequence (see Fig. 11.5).

basic (igneous rock): without quartz, formed from a magma that is undersaturated with silica; the ferromagnesian minerals are typically pyroxene and olivine.

basin (sedimentary): a depression on the Earth's surface that either contains, or is capable of containing, sediments; it may be of any size or shape, and is normally produced by tectonic activity (e.g. extension).

batholith: a deep-seated igneous body with a very large horizontal extent and great thickness, often with no determinable floor.

Benioff zone: inclined zone defined by **earthquake foci** used to indicate the presence of a **subduction zone** (see Fig. 6.6).

biostratigraphic hiatus: a gap in the fossil record that does not correspond to a recognizable gap in the stratigraphic record.

blind thrust: thrust whose **tip line** does not reach the erosion surface (see Fig. 8.12).

boudin, boudinage: a sausage-like shape produced as the result of the stretching of a layer; the process resulting in boudins .

branch line: line of intersection of a splay thrust with the main thrust (see Fig. 8.12).

brittle (behaviour): failure (fracture) after no, or very little, **plastic** or **viscous** deformation when a stress is applied.

buckling: folding process involving compression acting approximately parallel to the folded layers.

C

calc-alkaline: igneous rocks with a relatively high calcium to alkali ratio, characterized by the dominance of plagioclase rather than alkali feldspar.

Caledonian (orogeny), **Caledonides**: a series of orogenic episodes affecting most of the British Isles together with western Norway and eastern Greenland during Lower Palaeozoic time; the belt affected by this orogeny.

carbonates: general term for sedimentary strata dominated by carbonate minerals such as limestone and dolomite.

cataclasis, cataclastic: deformation dominated by fracturing.

catastrophism: early theory proposing that a single catastrophic process (such as the biblical flood) was responsible for the formation of rocks.

cephalopod: member of the Class Cephalopoda: marine molluscs, which include squids, octopuses and goniatites (see Fig. 3.8).

channel flow: gravity-induced lateral flow of a warm **ductile** layer in the lower crust of a thickened orogen, which is sandwiched between two stronger layers and squeezed towards the sides of the orogenic belt.

chert: rock consisting of microcrystalline silica, in the form of sedimentary layers or as nodules within limestones; formed either by the accumulation of siliceous microfossils or chemically, by precipitation.

clastic (sedimentary rock): resulting from the disintegration of older rocks.

clinoform: shape of a 'complete' bedding surface – i.e. followed from its proximal to its distal end (see Fig. 11.8).

co-axial strain: progressive strain in which the orientations of the **principal strain axes** remain unchanged throughout the deformation; also known as **pure shear** (see Fig. 5.3).

competent: (of a bed or layer of rock) stronger when deformed, compared to adjoining beds or layers.

concentric (fold): where the folded layers form arcs of circles about a common centre.

conservative (plate boundary): characterized by sideways motion against the adjoining plate along a **transform fault** (see Fig. 6.13).

constructive (plate boundary): where new plate is created by divergent motion of two adjacent plates, marked by an ocean ridge or continental rift (see Fig. 6.13).

continent: in a geological context, the landmass of a continent plus adjacent sea bed underlain by **continental crust**, including the **continental shelf** and **continental slope**.

continental crust: that part of the crust underlying a continent, with variable composition corresponding, on average, to a mixture of granite and basalt.

continental platform: the stable part of a continent, external to any orogenic activity.

continental shelf: that part of the continental margin covered by relatively shallow seas.

continental slope: zone between the **continental shelf** and the deep ocean.

convection current: a pattern of flow in liquid or solid material driven by a temperature difference; this produces a density imbalance which provides the force necessary to generate the flow.

Cordilleran (belt): orogenic belt situated in western North America resulting from a series of **subduction**-related events and collision of **terranes** during the Mesozoic and Cenozoic Eras.

craton: the stable part of a continental interior, unaffected by contemporary orogenic activity.

crenulation cleavage: type of foliation produced by sets of closely spaced crenulations (microfolds), of the order of millimetres or less in width; when viewed at outcrop scale, it has the appearance of a set of bands that cut across the original layering (see Fig. 5.14).

culmination: of a thrust complex (e.g. a **duplex**) where a series of thrust sheets is locally thicker, such that the topmost thrust bulges upwards and may be exposed by erosion as a tectonic **window**.

cumulate: ultrabasic igneous rock interpreted as a concentration of early-formed crystals deposited from a melt, from which the remaining material has been extracted (see Figs 7.7, 7.8).

cut-off (line): intersection of a **thrust** surface with a particular stratigraphic horizon .

cyclothem: sequence of sedimentary beds deposited as part of a single cycle of sedimentation; used to describe short-lived sequences such as those found typically in Carboniferous coal-bearing strata (see Fig. 11.10).

D

dacite: fine-grained, intermediate, **calc-alkaline** igneous rock.

décollement: see **detachment**.

deep-focus (earthquake): originating at a depth of below 300 km.

destructive plate boundary: a plate boundary characterized by the convergent movement of adjoining plates, and by the destruction of oceanic plate or the collision of continents (see Fig. 6.13).

detachment horizon (or surface): weak surface or layer, usually parallel to bedding, used as a **thrust** plane or surface of discontinuity between two different folding modes (see Fig. 10.2).

dextral (of a fault, shear zone, etc.): where the opposite side of a plane or zone moves to the right, as seen by an observer on one side.

diachronous: of a geological event, or events (e.g. sedimentation, volcanicity): not at the same time; spanning a period of time.

diapir, diapirism: body of rock, typically with a shape like an inverted tear drop, that has risen through the Earth's crust due to its buoyancy (see Fig. 12.5); process of emplacement of a diapir.

diorite: coarse-grained **intermediate igneous rock**.

disconformity: a stratigraphic surface, corresponding to a recognizable time interval, where the strata above and below are conformable (see Fig. 11.1).

downlap: where an initially inclined stratal surface terminates down-dip against the stratal surface beneath (see Fig. 11.5).

ductile: process in which failure (fracture) in a material occurs only after considerable **plastic** or **viscous** deformation when a stress is applied.

dunite: ultrabasic igneous rock composed mainly of olivine.

duplex: structure formed by a **imbricate** set of thrusts enclosed above by a **roof thrust** and beneath by a **floor thrust** (see Figs 8.13, 8.14).

dynamic: process governed by forces or stresses.

E

eclogite (facies): metamorphic assemblage formed under exceptionally high-pressure conditions.

elastic (strain): where the strained body quickly returns to its unstrained shape on removal of the applied stress; i.e. 'temporary' strain.

elongation lineation: linear **fabric** caused by the alignment of elongate objects (e.g. fossils or **porphyroblasts**) in a deformed rock (see Fig. 5.8).

erosional truncation: lateral termination of a sedimentary layer by erosion.

eugeosyncline: term formerly used to describe a type of depositional assemblage within an orogenic belt containing thick sequences of predominantly **clastic** deposits together with volcanics and, importantly, **ophiolites**; now regarded as the product of a combination of deep ocean, ocean trench and **continental slope** environments.

eustasy, eustatic: a global uniform change in sea level (i.e. not due to a local tectonic cause).

evaporite: a sedimentary deposit formed by the evaporation of warm shallow seas containing salts (especially sodium chloride) in solution.

extension: deformation process tending to produce elongation in a body.

F

fabric: a set of new structures, or a texture, produced in a rock as a result of deformation – e.g. **foliation** or **lineation**.

fault-plane solution: process in which the direction and shear sense of a fault displacement is calculated by analysing the first motion of the associated earthquake.

flat: (of a **thrust**): that portion which follows bedding, as distinct from a **ramp**, which cuts through bedding (see Fig. 8.13).

Flinn diagram: graph describing the shape of a deformed object, where the ratio of the major to intermediate strain axes is plotted against that of the intermediate to minor axes (see Fig. 5.5).

flood basalt: see **plateau basalt**.

focal mechanism: see fault-plane solution.

focus: (of an earthquake) the location of the origin.

foliation: a set of planar structures produced in rock as a result of deformation.

fore-arc: the region of an island arc system between the volcanic arc and the ocean trench.

foredeep (basin): a sedimentary basin resulting from the depression of the continental crust due to the load of the rising orogenic belt; it contains a thick sequence of predominantly **clastic** sediments derived from the erosion of the main mountain range.

foreland: that part of the **continental crust** lying immediately adjacent to an orogenic belt and which has not been significantly affected by it.

foreland basin: see **foredeep**.

formation: a set of sedimentary strata that have some feature(s) in common that make(s) it a recognizable stratigraphic unit.

G

gneiss: a coarse-grained metamorphic rock characterized by the segregation of light-coloured minerals such as quartz and feldspar into bands or lenses separated by dark minerals such as micas or hornblende.

geoid: the shape of the Earth's surface, measured as if at sea level.

geosyncline: obsolete term describing an elongate tectonic basin or trough, which, when filled with sediments, subsequently becomes part of an orogenic belt.

glacial rebound: gravitational response to the removal of the load of an ice sheet, resulting in the slow rise of the land surface.

Gondwana: a supercontinent that existed during Palaeozoic time, consisting of the continents of South America, Africa, India, Antarctica and Australia (see Fig. 3.10).

graben: down-faulted block bounded by **normal faults**.

granite: coarse-grained **acid** igneous rock containing quartz, feldspar and either mica or hornblende.

granitization: obsolete term to describe a process whereby sedimentary rock of a suitable composition is transformed by a metasomatic process into granite.

granophyre: fine- to medium-grained acid igneous rock, typically containing larger feldspar crystals in a groundmass of intergrown potassium feldspar and quartz.

granulite (facies): a metamorphic assemblage formed under high pressure and temperature.

gravity sliding (or gliding): movement of a body of rock down a slope due only to gravitational force (see Fig. 12.3).

gravity spreading: lateral spreading of the over-thickened crust of an orogen under gravity, leading to a reduction of crustal thickness; also known as '**orogen collapse**' (see Fig. 12.13).

greenschist (facies): a metamorphic assemblage formed under low temperature and pressure.

greywacke: an unsorted sandstone composed of a variety of minerals; typically formed by deposition from a **turbidity current**.

guyot: flat-topped submarine volcanic mountain, eroded to sea level and in some cases subsequently depressed as the ocean floor subsides (see Fig. 6.2).

H

halite: rock salt (sodium chloride).

harzburgite: ultrabasic igneous rock composed mainly of olivine and pyroxene.

Hercynian (orogeny): late Palaeozoic to early Mesozoic orogeny affecting much of Europe.

heterogeneous deformation, strain: where the amount of strain is variable throughout the strained body; straight lines become curved and parallel lines become non-parallel.

hiatus: in stratigraphy, a gap in a chronological sequence.

highstand systems tract: sedimentary sequence developed on the **continental platform** and **continental slope** during a period of high relative sea level (see Fig. 11.8).

hinterland (sedimentary) sequence: non-marine deposits laid down entirely landward of the coast, and not directly influenced by sea-level changes.

homogeneous deformation, strain: where the amount of strain does not vary throughout the strained body; straight lines stay straight and parallel lines stay parallel.

horse: structure formed by a pair of thrusts that diverge, then meet again at a higher level, e.g. within a **duplex** (see Fig. 8.13).

horst: uplifted block bounded by **normal faults**.

hot-spot: a part of the Earth's crust exhibiting unusually high heat flow and volcanicity, either within a plate (as in Hawaii) or on a **constructive** plate boundary (as in Iceland – see Fig. 10.6).

I

imbricate (structure): where the same sequence of strata is repeated many times in successive thrust slices (e.g. see Fig. 8.13).

intermediate igneous rock: an igneous rock intermediate in composition between **acid** and **basic**.

intraplate: relating to the interior of a tectonic plate (rather than at a plate boundary).

irrotational (strain): deformation in which the strain axes do not change their relative position during the strain process; see also **co-axial strain**.

island arc: a submerged arcuate mountain range, typically volcanic, situated alongside a **subduction zone** at a **destructive** plate boundary (see Fig. 6.1).

isostasy: the theory or state of general gravitational equilibrium at the Earth's surface, in which topographic variations are balanced by density variations beneath (see Fig. 12.1).

K, L

kinematic: process governed by movement.

lapout: lateral termination of a stratum at its original depositional limit (see Fig. 11.8).

Laurentia: a continent consisting of most of North America, Greenland and NW Scotland that existed during the Lower Palaeozoic prior to the **Caledonian orogeny**.

leading cut-off: line where a thrust surface first meets a stratigraphic horizon.

Lewisian (complex): Archaean to Late Proterozoic basement, mostly composed of **gneisses**, of the NW **foreland** of the Scottish **Caledonides**.

lherzolite: coarse-grained ultrabasic igneous rock; a type of **peridotite** composed of olivine plus lesser amounts of orthopyroxene and clinopyroxene, together with minor quantities of plagioclase and spinel; capable of partial melting to yield basalt.

lineation: a set of linear structures (i.e. a linear **fabric**) resulting from deformation.

listric fault: a curved **normal fault** whose inclination decreases downwards (see Fig. 12.7).

lithification: process whereby sediment is consolidated to become solid rock.

lithosphere: the strong upper layer of the Earth, with an average thickness of about 100 km, including the crust and part of the mantle; it consists of a number of plates that move over the weaker **asthenosphere** beneath (see Fig. 6.10).

lowstand systems tract: sedimentary sequence developed on the **continental slope** and basin floor during a period of low relative sea level (see Fig. 11.8).

low-velocity layer: region of the Earth's mantle at depths of *c.*100–200 km through which seismic waves travel more slowly than would be expected if it were composed of normal mantle material; it is regarded as a zone of relative weakness; also known as the **asthenosphere**.

M

magnesio-wüstite: a magnesium-iron oxide mineral (MgFeO) with a cubic crystal structure; wüstite contains less oxygen than other iron ores and is often found in meteorites.

mantle drag: the force exerted on the base of a tectonic plate by the frictional drag of the underlying mantle.

mantle plume: a column of rising hot mantle material inferred to explain **hot-spots**.

maritime (sedimentary) sequence: sequence of coastal and/or marine sedimentary deposits where the coastal element is controlled by the position of sea level.

mélange: rock body, which may be of either sedimentary or tectonic origin, containing disrupted fragments of pre-existing rocks, metres to hundreds of metres in size.

metamorphic core complex: interior part of an **orogenic belt** typified by igneous and metamorphic rocks originally formed at deep-crustal levels and brought to the surface as a result of the orogeny.

migmatite: type of metamorphic rock where a (usually) **gneissose** host rock is cut by numerous veins and/or bands of **granitic** or quartzo-feldspathic material.

miogeosyncline: obsolete term formerly used for a type of depositional assemblage within an orogenic belt containing a thick sequence of predominantly **clastic** deposits; now regarded as the product of the **continental slope** environment.

Mohorovičić discontinuity (Moho): the base of the crust, determined by seismic reflection.

Moine thrust (zone): major **thrust** (zone of thrusts) in NW Scotland defining the western boundary of the Caledonian orogenic belt (see Fig. 8.11).

mylonite: a fine-grained fault rock formed at depth under metamorphic conditions and typically showing a regular fine banding; associated with major **thrust** zones.

N

nappe: a large sheet of rock, many kilometres in extent, resting on a basal fault, typically a **thrust** (e.g. see Fig. 8.11).

Neptunism: early theory which held that all rocks (including igneous rocks) had formed beneath the sea.

normal fault: a dip-slip fault whose upper side has moved down the fault plane (e.g. see Fig. 12.7).

O

obduction: process in which part of the oceanic **lithosphere** is detached and thrust over **continental crust** (e.g. see Fig. 7.4).

oblate strain (ellipsoid): three-dimensional strain characterized by the minimum strain axis being much shorter than the maximum and intermediate strain axes, which are equal or near equal; 'flattening' strain (see Fig. 5.2).

ocean-floor spreading: process of formation of new **oceanic crust** by the injection of magma at an **ocean ridge** (see Fig. 7.3).

ocean ridge: long, submerged oceanic mountain range, site of a **constructive plate boundary** (see Figs. 6.1, 6.13).

ocean trench: deep marine trough, site of a **subduction** zone (see Figs. 6.1, 6.13).

oceanic plateau: an especially thick and buoyant piece of **oceanic crust** that is less dense than normal and may become accreted to the upper **plate** of a **subduction zone**.

onlap: type of **baselap** where an originally horizontal stratum laps out against a surface with a greater inclination (see Fig. 11.8).

oolite: sedimentary rock composed of small spherical carbonate concretions, often with nuclei of sand grains, formed in shallow warm seas.

ophiolite: a sequence of rock types interpreted as pieces of oceanic crust; a 'complete' sequence includes **ultrabasic** material assumed to be from the uppermost mantle, sheet-like basic and ultrabasic intrusions, a **sheeted dyke** layer, basalt **pillow lavas** and ocean-floor sediments (see Fig. 7.7).

orogen collapse: (also known as 'orogen spreading'): process of lateral spreading of an over-thickened **orogen** (see Fig. 12.13).

orogenic belt, or **orogen**: part of the Earth's crust, typically a linear zone, that has undergone **orogenesis** (mountain building) as a result of **plate** convergence, causing crustal thickening, uplift and the formation of mountains.

overthrust (fault): see **thrust**.

P

palaeomagnetism: the study of the magnetic properties of rocks; principally to determine the orientation of their magnetic latitude and pole position at some point in the geological past.

Pangaea: the **supercontinent**, consisting of the whole continental landmass, which existed during much of Upper Palaeozoic time (see Figure 3.2).

paracycle: a (usually local) sedimentary sequence caused by a rapid rise of sea level followed by a **stillstand** contained within a larger cycle of relative sea-level change (see Fig. 11.10).

parallel fold: a fold in which the folded layers are of approximately equal thickness throughout.

passive continental margin: corresponding to a now inactive **constructive plate boundary**.

pericline: fold which varies in height along its length such that the amplitude decreases to zero at each end, and may be either anticlinal or synclinal (see Fig. 8.3).

peridotite: an ultrabasic rock containing a high proportion of pyroxene and olivine.

perovskite: calcium-titanium oxide mineral with a cubic crystal structure.

pillow lava: lava deposited in water, especially on the ocean floor, and characterized by structures resembling pillows and tubes (see Fig. 7.1B).

pin line: reference line from which bed length is measured in constructing a **balanced section**.

plagioclase: group of light-coloured, alumino-silicate feldspar minerals containing variable amounts of sodium and calcium.

plane strain: three-dimensional **strain** characterized by the intermediate **principal strain axis** remaining unchanged (see Figure 5.2).

plastic (strain): permanent **strain**, remaining after the deforming stress is removed.

plate: a relatively stable piece of the **lithosphere** that moves independently of adjoining plates.

plate tectonics: the theory that ascribes tectonic processes to the relative movement of the **lithosphere plates**.

plateau basalt: collective term for a voluminous set of (typically tholeiitic) basaltic lava flows covering large areas to form vast plateaux.

plume (mantle): column of hot rising mantle material thought to be responsible for **hot-spots** on the Earth's surface.

plunge: angle and/or direction of inclination of a fold axis or **lineation**, measured from the horizontal.

pluton: an igneous body of large dimensions, both horizontally and vertically.

Plutonist: obsolete term describing early geologists who, following James Hutton, believed that crystalline rocks such as granite or basalt that are now known to be igneous were formed by solidification of magma.

pohlflucht: (literally, 'flight from the poles'): obsolete theory ascribing movement of the continents (continental drift) to a component of the Earth's centrifugal force (now known to be far too weak) acting from the poles towards the Equator.

poise: cgs unit of dynamic **viscosity**, measured by the force required to produce a flow velocity of 1 cm/sec through an area of 1 cm^2 of a fluid.

polar wandering: term used to describe either the apparent movement through time of a former pole position, or the **precession** of the actual pole position.

pole (of rotation): point where a rotation axis penetrates the Earth's surface (see Fig. 6.12).

porphyroblast: a large crystal in a metamorphic rock.

precession: the slow movement of the axis of a rotating body about another axis (in the case of the Earth, the rotational axis moves around its mean position).

preferred orientation: where a set of objects (e.g. crystals) in a rock have, or tend towards, a parallel orientation **fabric** caused by deformation.

primary (P-) waves: the first set of **earthquake waves** to arrive at a recording station; they propagate by a process of alternate expansion and compression similar to sound waves.

principal strain (axes): directions of the mutually perpendicular greatest, least and intermediate **strain** directions in a deformed body (see Fig. 5.2).

prolate strain (ellipsoid): three-dimensional **strain** characterized by the maximum **principal strain axis** being much larger than the intermediate and minimum strain axes, which are equal or nearly equal; i.e. elongation strain (see Figure 5.2B).

pure shear: see **co-axial strain**.

pyrolite: theoretical rock composed of 3 parts dunite to one part tholeiitic basalt thought to represent the composition of the mantle.

R

radiolarian chert: a **siliceous** deposit formed from the skeletons of *radiolaria* (unicellular marine organisms) and found in deep-ocean environments.

raised beach: a former beach, above the present shore line, formed at a time of higher sea level (see Figure 12.2).

ramp: section of a **thrust fault** that cuts up through bedding (see Figure 8.13C).

reflection seismology: technique of studying structures at depth by analysing the reflection pattern of artificially generated seismic waves.

regional metamorphism: metamorphism generated by regional changes in temperature and/or pressure.

rhyolite: fine-grained acid igneous rock.

rollover: antiformal structure formed on the upper side of a large normal fault due to the gravitational effect of the fault displacement (see Fig. 12.8).

roof thrust: the top thrust of a **duplex** (see Fig. 8.13B).

root zone: where the main thrusts (including the **sole thrust**) of a **thrust** complex eventually descend into the lower crust or mantle

rotational strain: progressive strain in which the **principal strain axes** rotate during the deformation; also known as **simple shear** or 'non-co-axial' strain (see Fig. 5.12).

S

salt diapir: intrusive salt body (see Figure 12.5).

salt dome: a dome-shaped structure formed by the upward migration of salt under gravitational pressure from its source layer (see Figure 12.5).

schist: a medium- or coarse-grained metamorphic rock characterized by the parallel alignment of platy minerals such as chlorite or mica.

schistosity: the **foliation** characteristic of a **schist**.

schuppen (structure): see **imbricate** structure.

sea-floor spreading: the theory of the generation of **ocean crust** at **ocean ridges** and its destruction at **ocean trenches** (see Figure 6.3).

sechron: total interval of time during which a **stratigraphic sequence** is deposited.

secondary (S-) waves: the second set of **earthquake waves** to arrive at a recording station; they are propagated by a process of lateral vibrations (i.e. **shear**).

seismic, seismicity: relating to **earthquake** activity.

seismic section: cross-section through the crust obtained from artificial seismic sources to elucidate sub-surface structure, usually for prospecting purposes (see Figs. 11.6, 7).

seismic (wave) tomography: method of analysing seismic velocity variations at depth by comparing arrival times of earthquake waves travelling in different directions through the same volume of the mantle in order to detect areas of anomalous velocity.

sequence stratigraphy: the study of sets of sedimentary strata based on the nature of their unconformable boundaries and their geometric relationships with the stratal surfaces.

serpentine: collective name for a group of hydrated silicate minerals such as chlorite that occur as alteration products replacing ferromagnesian minerals such as pyroxene, olivine or hornblende.

shallow-focus (earthquake): originating at a depth of 0 to 60 km below the surface.

shear: (applied to a plane, e.g. a fracture or fault): process resulting from an oblique **stress** tending to produce opposed directions of movement on either side of the plane.

shear strain: rotational **strain**; i.e. where the strain axes progressively rotate during the deformation.

shear zone: a zone of **ductile** deformation between two rock masses moving in opposite directions; the equivalent of a fault at depth (see Fig. 5.9).

sheeted dyke (layer, complex): part of the **oceanic crust** composed entirely of steep **dykes** that have acted as feeders to the **basalt** layer above (see Figs 7.7, 7.8).

sial: obsolete term describing a crustal layer composed predominantly of material with a broadly granitic composition.

sigma structure: asymmetric shape of a deformed **porphyroblast**, named after the Greek letter σ, used to determine the **shear** sense in a deformed **schist** or **gneiss** (see Fig. 5.14C).

sima: obsolete term describing a crustal layer composed predominantly of basic igneous rocks such as basalt.

similar (fold): where the folded layers vary in thickness but where each layer has the same shape.

simple shear: see **rotational strain**.

sinistral: (of a fault, **shear zone** etc.): where the opposite side of the structure moves to the left as seen by an observer on one side.

slab-pull (force): the force exerted on a **tectonic plate** by the gravitational pull of a sinking **lithosphere** slab at a **subduction zone** (see Fig. 9.6).

slab roll-back: process where the position of a trench (i.e. **subduction zone**) retreats ocean-wards due to the **slab-pull** effect (see Fig. 9.6).

slip vector: of an earthquake, the direction of the first motion on the fault plane.

sole thrust, fault: basal thrust (or fault) of a fault system (e.g. see Fig. 8.11).

splay (fault): subsidiary fault formed at an angle to an existing fault (see Fig. 8.12).

spinel: group of oxide minerals with a simple cubic crystal structure (e.g. magnetite $-Fe_3O_4$).

Steinmann trinity: the three components of an **ophiolite** complex according to Gustav Steinmann: **serpentinized** peridotite, **pillow basalt** and deep-marine bedded **chert**.

stillstand: period when sea level maintains an apparently constant position relative to the deposition surface (see Fig. 11.10).

strain: change in shape produced by a **stress** system.

strain ellipsoid: a geometrical device to illustrate the three-dimensional variation of **strain** (as if resulting from deformation of a sphere) in a deformed body (see Fig. 5.2).

stress: a pair of equal and opposite forces acting on unit area of a surface.

strike-slip (fault etc.): where the movement has taken place horizontally along the fault plane (see **wrench fault**).

structural truncation: lateral termination of a sedimentary layer by structural disruption, e.g. by a fault.

subduction: the process whereby an oceanic **plate** descends into the **mantle** along a subduction zone; (see Fig. 6.10).

subduction suction: force exerted on the upper plate of a **subduction zone** caused by **slab-pull**; also called 'trench suction' (see Fig. 9.6).

supercontinent: a large continental mass consisting of several components that previously, or subsequently, were themselves continents.

superposition (law of): younger strata always lie above older.

surface waves: set of **earthquake waves** that travel around the surface of the Earth's crust and arrive at a recording station after the **primary** and **secondary waves**.

suture: the boundary between two continental **plates** brought together during collisional **orogeny**.

synform: a fold that closes downwards.

T

terrane: a piece of the Earth's crust, smaller in scale than a **plate**, and now part of an **orogenic belt**, which exhibits a different structural history from its neighbours and contains palaeomagnetic or fossil evidence of derivation from some distance away.

Tethys ocean: the ocean that separated **Laurasia** and **Gondwana** during Upper Palaeozoic and early Mesozoic time.

thick-skinned: of a fault complex: where the faults penetrate to deep crustal levels.

thin-skinned: of a fault complex: confined to a relatively thin upper layer of the crust, resting on a basal **detachment horizon** or **sole thrust** (e.g. see Fig. 8.7).

tholeiite: type of silica-rich **basalt**.

thrust (fault): a fault that places older rocks above younger as a result of compression; it is generally gently inclined ($<45°$) although it may become steepened by subsequent movements (see Fig. 8.13).

thrust sheet: body of rock resting on a **thrust**; see also **nappe**.

tip line: the margin of a thrust surface (see Fig. 8.12).

tonalite: coarse-grained **acid** igneous rock characterized by an absence of alkali-feldspar.

toplap: **lapout** at the upper boundary of a depositional sequence (see Fig. 11.5).

trailing cut-off: line where a **thrust** surface leaves a stratigraphic horizon.

transcurrent (fault): see **wrench** fault.

transform fault: a fault that forms part of a **plate** boundary where the plates on each side move in opposite directions, parallel to the trend of the fault (see Figure 6.7).

transgression: a rise in sea level such that the shore-line progressively encroaches over the land.

transgressive systems tract: sedimentary sequence developed during a marine **transgression** (see Fig. 11.9A).

trench roll-back: see **slab roll-back**.

triple junction: meeting point of three **constructive plate boundaries**, one of which may be inactive (e.g. see Fig. 6.15).

turbidity current: a water current generated by gravity-induced flow, carrying large quantities of sediment of varying coarseness in suspension.

U, V, W

uniformitarianism: early theory proposing that the processes governing the formation of rocks have always been similar to those operating at present.

viscosity: the quality of stickiness or fluidity of a material, measured by its rate of flow.

viscous: of a material, semi-fluid; with a slow rate of flow.

Wheeler diagram: chart portraying stratigraphic sequences in terms of time equivalence rather than lithological similarity (see **sequence stratigraphy**) (see Fig. 11.2).

window (tectonic): topographic area where erosion has cut through a cover sequence to reveal the structure at a lower level.

wrench fault: a fault where the displacement is horizontal and parallel to the trend of the fault; see also **strike-slip** fault (see Fig. 6.7).

References

Chapter 1

Cuvier, George (Baron) (1821) *Essay on the theory of the Earth*. Edinburgh, Blackwood; London, John Murray.

De Saussure, H.B. (1779–1786) *Voyages dans les Alpes*. Neuchatel, Fauche.

Hutton, James (1788) *Theory of the Earth; or an investigation of the laws observable in the composition, dissolution and restoration of land upon the globe*. Transactions of the Royal Society of Edinburgh **1**, 209–304.

— (1795) *Theory of the earth; with proofs and illustrations*. Edinburgh, Creech.

Lyell, Charles (1830–1833) *Principles of Geology, being an attempt to explain the former changes of the Earth's surface, by reference to causes now in operation*. London, John Murray.

Lyell, Charles (1838–1865) *Elements of Geology*. London, John Murray.

Werner, Abraham (1774) *Von der äusserlichen Kennzeichen der Fossilien* (on the origin of the external characteristics of fossils). Leipzig, Elegfried Lebrecht.

Chapter 2

Darwin, Charles (1859) *On the Origin of Species by Means of Natural Selection, or the Preservation of Favoured Races in the Struggle for Life*. London, John Murray.

Darwin, Charles (1871) *The Descent of Man, and Selection in Relation to Sex*. London, John Murray.

Darwin, Charles (1881) *The formation of vegetable mould under the action of worms*. London, John Murray.

Darwin, Charles and Alfred Russel Wallace (1858) On the Tendency of Species to form Varieties; and on the Perpetuation of Varieties and Species by Natural Means of Selection. *Journal of the Proceedings of the Linnean Society of London. Zoology* **3** (9) 46–50.

Dawkins, Richard (1990) *The blind watchmaker*. London, Penguin Science.

De Vries, Hugo (1901–03) *Die mutationstheorie. Versuche und beobachtungen über die entstehung von arten im pflanzenreich*. Leipzig, Veit & Co.

Lamarck, Jean-Baptiste (1809) *Philosophie zoologique, ou exposition des considérations relatives à l'histoire naturelle des animaux*. Paris, Dentu et L'auteur.

Malthus, Thomas (1798) *An essay on the principle of population as it affects the future improvement of society with remarks on the speculations of Mr. Godwin, M. Condorcet, and other writers*. Published anonymously.

Mendel, Gregor (1866) *Versuche über Pflantzen-Hybriden* (experiments on plant hybridization). Brünn, Verhandlungen des naturforschunder Vereines.

Paley, William (1802) *Natural theologie or evidence of the existence and attributes of the Deity*. London, R. Faulder; Philadelphia, John Morgan.

Wallace, Alfred Russsel (1855) On the law which has regulated the introduction of new species. *Annals and magazine of Natural History*.

Watson, J.D. and Crick, F.H. (1953) Genetical implications of the structure of deoxyribonucleic acid. *Nature* **171** (4361): 964–7.

Chapter 3

Argand, E. (1924) La tectonique de l'Asie. *Comptes rendus du congrès géologique internationale, Belgique*, 1922, 171–372.

Bertrand, M.A. (1884) Rapports de structure des Alpes de Glaris et du bassin houiller du Nord. *Bulletin de la Societé Géologique de France, 3ème Séries* **12**, 318–330.

Dana, J.D. (1863) *Manual of Geology*. Philadelphia, Theodore Bliss; London, Trübner & Co.

De Beaumont, L. Elie (1852) *Notice sur le système des montagnes*. Paris.

Du Toit, A.L. (1921) The Carboniferous glaciation of South Africa. *Transactions of the Geological Society of South Africa* **24**, 188–217.

— (1927) A geological comparison of South America and South Africa. *Carnegie Institute, Washington* **381**, 1–157.

— (1937) *Our Wandering Continents. An Hypothesis of Continental Drifting*. London, Oliver & Boyd.

Epstein, P.S. (1921) Uber die Pohlflucht der Continente. *Die Naturwissenschaften* **9**, 499–502.

Geikie, A. (1893) *Textbook of Geology*, (3rd edn). London, McMillan & Co.

Jeffreys, H. (1924) *The Earth, its origin, history and physical constitution*. Cambridge University Press.

— (1935) *Earthquakes and Mountains*. Cambridge University Press.

Kirsch, G. (1928) *Geologie und Radioaktivität*. Vienna and Berlin, Springer.

Runcorn, S.K. (1962) Palaeomagnetic evidence for continental drift and its geophysical cause. In: Runcorn, S.K. (ed.) *Continental drift*. New York, London, Academic Press.

Schweydar, W. (1921) Bemerkungen zu Wegeners Hypothese der Verschiebung der Continente. *Zeitschrift der Gesellschaft für Erdkunde zu Berlin*, 120–125.

Şengör, A.M.C. (1982) Classical theories of orogenesis. In Miyashiro, A., Aki, K. & Şengör, A.M.C. *Orogeny*, Chichester, John Wiley.

Suess, E. (1875) *Die Enstehung der Alpen*, Vienna, Braumüller.

— (1885–1901) *Das Antlitz der Erde*. Leipzig, G. Freytag.

Taylor, F.B. (1910) Bearing of the Tertiary mountain belt on the origin of the Earth's plan. *Bulletin of the Geological Society of America* **21** (2), 179–226.

Wegener, A. (1912) Die Enstehung der Kontinente. *Geologische Rundschau* **3**, 276–292.

— (1915–1929) *Die Enstehung der Continente und Ozeane*. Braunschweig, F. Vieweg und Sohn.

— (1924) *The Origin of Continents and Oceans*. London, Methuen & Co.

Chapter 4

Becquerel, H. (1896) Sur les radiations émises par phosphorescence. *Comptes Rendus* **122**, 501–503.

Curie, P. and Laborde, A. (1903) Sur la chaleur dégagée spontanément par les sels de radium. *Comptes Rendus Hebdomadaires des Sciences de l'Academie des Sciences* **136**, 673–675.

Holmes, A. (1913) *The age of the Earth*, London & New York, Harper Bros.

— (1929) Radioactivity and earth movements. *Transactions of the Geological Society, Glasgow* **18**, 559–606.

— (1944) *Principles of Physical Geology*. Edinburgh, Thomas Nelson.

Jeffreys, H. (1925) The cooling of the Earth. *Nature* (London) **115**, 876–878.

— and Bullen, K.B. (1935) Time of transmission of earthquake waves. *Publications du Bureau Central Seismologique International* AII. 202pp.

Joly, J. (1899) An estimate of the geological age of the Earth. *Scientific Transactions of the Royal Dublin Society* **7**, 23–66.

— (1909) *Radioactivity and geology: an account of the influence of radioactive energy on terrestrial history.* New York, Van Norstrand.

— (1924) *Radioactivity and the surface history of the Earth.* Oxford University Press.

Lehmann, I. (1936) *Publications du Bureau Central Seismologique International* **A14**, 87–115.

Mohorovičić, A. (1909) Das Beben vom 8 x 1909. *Jahrbuch des Meteorologischen Observatoriums in Zagreb* **9**, 1–63.

Oldham, R.D. (1906) The constitution of the interior of the Earth, as revealed by earthquakes. *Quarterly Journal of the Geological Society, London* **62**, 456–475.

Rutherford, E. and Soddy, F. (1902) The cause and nature of radioactivity. *Philosophical Magazine* **4**, 370–396, 569–585.

Thomson, W. (1864) On the secular cooling of the Earth. *Transactions of the Royal Society of Edinburgh* **23**, 167–169.

Vening Meinesz, F.A. (1962) Thermal convection in the Earth's mantle. In S.K. Runcorn (ed.) *Continental drift.* London, Academic Press.

Chapter 5

Escher, A. and Watterson, J. (1974) Stretching fabrics, folds and crustal shortening. *Tectonophysics* **22**, 223–231.

Flinn, D. (1962) On folding during three-dimensional progressive deformation. *Quarterly Journal of the Geological Society, London* **114**, 385–433.

Ramberg, H. (1959) Evolution of ptygmatic folding. *Norsk Geologisk Tidsskrift* **39**, 99–152.

Ramsay, J.G. (1967) *Folding and fracturing of rocks.* London, McGraw-Hill.

— (1980) Shear zone geometry: a review. *Journal of Structural Geology* **2**, 83–99.

— and Graham, R. H. (1970) Strain variation in shear belts. *Canadian Journal of Earth Science* **7**, 786–813.

Sander, B. (1930) *Gefügekunde der Gesteine.* Vienna, Springer.

Shackleton, R.M. and Ries, A.C. (1984) The relation between regionally consistent stretching lineations and plate motions. *Journal of Structural Geology* **6**, 111–117.

Skjernaa, L. (1980) Rotation and deformation of randomly oriented planes and linear structures in progressive simple shear. *Journal of Structural Geology* **2**, 101–109.

Watterson, J. (1968) Homogeneous deformation of the gneisses of Vesterland, South-west Greenland. *Meddeleser om Grønland* **175**, 6, 72pp.

Chapter 6

Barazangi, M. and Dorman, J. (1968) World seismicity map of ESSA Coast and Geodetic Survey epicenter data for 1961–1967. *Seismological Society of America Bulletin* **59**, 369–380.

Benioff, H. (1949) Seismic evidence for the fault origin of oceanic deeps. *Geological Society of America Bulletin* **60**, 1837–1866.

Hallam, A. (1973) *A revolution in the Earth Sciences: from continental drift to plate tectonics.* Oxford University Press.

Hess, H.H. (1962) History of ocean basins. In Engel, A.E.J. *et al.* (eds) (1962): *Petrologic studies: a volume in honor of A.F. Buddington.* Geological Society of America Boulder, Colorado.

Isacks, B., Oliver, J. and Sykes, L.R. (1968) Seismology and the new global tectonics. *Journal of Geophysical Research* **73**, 5855–5899.

Le Pichon, X. (1968) Sea-floor spreading and continental drift. *Journal of Geophysical Research* **73**, 3661–3697.

McKenzie, D.P. & Parker, R.L (1967) The North Pacific: an example of tectonics on a sphere. *Nature* (London) **216**, 1276–1279.

Morgan, W.J. (1968) Rises, trenches, great faults and crustal blocks. *Journal of Geophysical Research* **73**, 1959–1982.

Wilson, J.T. (1963) Hypothesis of Earth's behaviour. *Nature* **198**, 925–929.

— (1965) A new class of faults and their bearing on continental drift. *Nature* (London) **207**, 343–347.

Chapter 7

Anonymous (1972) Penrose Field Conference on ophiolites. *Geotimes* **17**, 24–25.

Bott, M.H.P. (1971) *The interior of the Earth*. London, Edward Arnold.

Brongniart, A. (1813) Essai de classification mineralogique des roches mélanges. *Journal des Mines* **34**, 190–199.

— (1821) Sur le gisement ou position relative des ophiolites, euphotides, jaspes etc. dans quelques parties des Apennines. *Annales des Mines, Paris* **6**, 177–238.

Cann, J.R. (1968) Geological processes at mid-ocean ridges. *Geophysical Journal, Royal Astronomical Society* **15**, 331–341.

Church, W.R. and Stevens, R.K. (1971) Early Palaeozoic ophiolites of the Newfoundland Appalachians as mantle–oceanic crust sequences. *Journal of Geophysical Research* **76**, 1460–1466.

Coleman, R.G. (1971) Plate tectonic emplacement of upper mantle peridotites along continental edges. *Journal of Geophysical Research* **76**, 1212–1222.

— (1977) *Ophiolites: ancient oceanic lithosphere*. Springer-Verlag.

Coney, P.J., Jones, D.L. and Monger, J.W.H. (1980) Cordilleran suspect terranes. *Nature* **288**, 329–333.

De Roever, W.P. (1957) Sind die alpintypen Peridotitmassen vielleicht tektonisch verfrachtete Bruchstüke der Peridotitschale? *Geologische Rundschau* **46**, 137–146. (trans.) 'Could the Alpine-type peridotite massifs be tectonic fragments of the peridotite layer?'

Dilek, Y. (2003) Ophiolite concept and its evolution. In Dilek, Y. and Newcomb, S. (eds) *Ophiolite concept and the evolution of geological thought*. Geological Society of America, Special Paper 373.

— and Newcomb, S. (eds) (2003) *Ophiolite concept and the evolution of geological thought*. Geological Society of America, Special Paper 373.

Ewing, M. (1965) The sediments of the Argentine basin. *Quarterly Journal, Royal Astronomical Society* **6**, 10–27.

Gass, I.G. (1968) Is the Troodos massif of Cyprus a fragment of Mesozoic ocean crust. *Nature* **220**, 39–42.

— (1980) The Troodos massif: its role in the unravelling of the ophiolite problem and its significance in the understanding of constructive plate margin processes. In Panayiotou, A. (ed.) *Ophiolites, Proceedings of the International Ophiolite Symposium, Cyprus, 1979*. Nicosia, Cyprus Geological Survey Department, 23–35.

— and Masson-Smith, D. (1963) The geology and gravity anomalies of the Troodos massif,

Cyprus. *Royal Society, London, Philosophical Transactions* **A225**, 417–467.

Glennie, K.W., Boeuf, M.G.A., Hughes-Clarke, M.W., Moody-Stuart, M., Pilaar, W.F.H. and Reinhardt, B.M. (1974) Geology of the Oman Mountains. *Koninklijk Nederlands geologisch mijnbouwkundig Genootschap, Verhandelingen* **31**. (3 vols.)

Gutenberg, B. (1959) *Physics of the Earth's interior.* New York and London, Academic Press.

Hess, H.H. (1955) Serpentines, orogeny and epeirogeny. In Poldervart, A. (ed.) *Crust of the Earth (A Symposium).* Geological Society of America Special Paper **62**, 391–407.

Kusznir, N.J. and Bott, M.H.P. (1976) A thermal study of the formation of oceanic crust. *Geophysical Journal, Royal Astronomical Society* **47**, 83–95.

Miyashiro, A. (1973) The Troodos complex was probably formed in an island arc. *Earth and Planetary Science Letters* **19**, 218–224.

— (1975) Classification, characteristics, and origin of ophiolites. *Journal of Geology* **83**, 249–281.

Moores, E.M. (1982) Origin and emplacement of ophiolites. *Reviews of Geophysics and Space Physics* **20**, 735–760.

— and Vine, F.J. (1971) The Troodos massif, Cyprus and other ophiolites as oceanic crust: evaluation and implications. *Royal Society, London, Philosophical Transactions* **268A**, 443–466.

Pearce, J.A. (1975) Basalt geochemistry used to investigate past tectonic environments on Cyprus. *Tectonophysics* **25**, 41–67.

Raff, A.D. and Mason, R.G. (1961) Magnetic survey off the coast of North America, 40°N latitude to 52°N. *Geological Society of America Bulletin* **72**, 1267–1270.

Steinmann, G. (1927) Die ophiolithischen Zonen in dem mediterranen Kettengebirge. *14th International Geological Congress, Madrid* **2**, 638–667.

Vine, F.J. and Matthews, D.H. (1963) Magnetic anomalies over ocean ridges. *Nature* **199**, 947–949.

Vuagnat, M. (1963) Remarques sur la trilogie serpentinites-gabbros-diabases dans le bassin de la Mediterranée occidentale. *Geologische Rundschau* **53**, 336–357.

Chapter 8

Anderson, E.M. (1942) *The dynamics of faulting and dyke formation: with applications to Britain.* Edinburgh, Oliver and Boyd.

Bally, A.W., Gordy, P.L. and Stewart, G.A. (1966) Structure, seismic data and orogenic evolution of the southern Canadian Rockies. *Canadian Association of Petroleum Geologists, Bulletin* **14**, 337–381.

Boyer, S.E. and Elliott, D. (1982) Thrust systems. *American Association of Petroleum Geologists, Bulletin* **66**, 1196–1230.

Butler, R.W.H. (1982) The terminology of structures in thrust belts. *Journal of Structural Geology* **4**, 239–245.

Dahlstrom, C.D.A. (1969) Balanced cross sections. *Canadian Journal of Earth Sciences* **6**, 743–757.

Elliott, D. and Johnson, M.R.W. (1980) Structural evolution in the northern part of the Moine Thrust Zone. *Royal Society of Edinburgh, Transactions (Earth Sciences)* **71**, 69–96.

Gwinn, V.E. (1964) Thin-skinned tectonics in the Plateau and Northwestern Valley and Ridge Provinces of the Central Appalachians. *Geological Society of America, Bulletin* **75**, 863–900.

King, P.B. (1951) *Tectonics of Middle North America.* Princeton, New Jersey, Princeton

University Press.

Laubscher, H.P. (1962) Die Zveiphasenhypothese der Jurafaltung. *Eclogae Geologicae Helveticae* **55**, 1–22.

— (1965) Ein kinematisches Modell der Jurafaltung. *Eclogae Geologicae Helveticae* **58**, 231–318.

— (1987) The 3D propagation of décollement in the Jura. In McClay, K.R. and Price, N.J. (eds) *Thrust and nappe tectonics*. Geological Society of London, Special Publications **9**, 311–318.

McClay, K.R. and Coward, M.P. (1981) The Moine thrust zone: an overview. In McClay, K.R. and Price, N.J. (eds) *Thrust and nappe tectonics*. Geological Society of London, Special Publications **9**, 241–260.

Peach, B.N., Horne, J., Gunn, W., Clough, C.T., Hinxman, L.W. & Teall, J.J.H. (1907) *The geological structure of the North-west Highlands of Scotland*. Memoirs of the Geological Survey of Great Britain.

Price, N.J. (1966) *Fault and joint development in brittle and semi-brittle rock*. London, Pergamon.

Ramsay, J.G. (1967) *Folding and fracturing of rocks*. New York, McGraw-Hill.

— (1981) Tectonics of the Helvetic Alps. In McClay, K.R. and Price, N.J. (eds) *Thrust and nappe tectonics*. Geological Society of London, Special Publications **9**, 293–309.

Suess, E. (1904) *Das Antlitz der Erde* (English translation) vol. 1. Oxford, Clarendon Press.

Chapter 9

Barazangi, M. and Isacks, B. (1971) Lateral variations of seismic wave attenuation in the upper mantle above the inclined earthquake zone of the Tonga island arc: deep anomaly in the upper mantle. *Journal of Geophysical Research* **76**, 8493–8516.

Bott, M.P. and Kusznir, N.J. (1984) Origin of tectonic stress in the lithosphere. *Tectonophysics* **105**, 1–14.

Cross, T.A. and Pilger, R.H. (1982) Controls of subduction geometry, location of magmatic arcs and tectonics of arcs and back-arc regions. *Geological Society of America, Bulletin* **93**, 545–562.

Dewey. J. (1980) Episodicity, sequence and style at convergent plate boundaries. *Geological Association of Canada, Special Paper* **20**, 553–574.

Elsasser, W.M. (1971) Sea-floor spreading as thermal convection. *Journal of Geophysical Research* **76**, (5), 1101–1112.

Forsyth, D. and Uyeda, S. (1975) On the relative importance of driving forces of plate motion. *Geophysical Journal, Royal Astronomical Society* **43**, 163–200.

Isacks, B. and Molnar, P. (1971) Mantle earthquake mechanisms and the sinking of the lithosphere. *Nature* **223**, 1121.

Karig, D.E. (1970) Ridges and basins of the Tonga-Kermadec island arc system. *Journal of geophysical research* **75**, (2), 239–254.

— (1971) Structural history of the Mariana Island arc system. *Geological Society of America, Bulletin* **82**, 323–344.

Molnar, P. and Atwater, T. (1978) Interarc spreading and cordilleran tectonics as alternates related to age of subducted oceanic lithosphere. *Earth and Planetary Science Letters* **41**, 330–340.

Oliver, J., Isacks, B., Barazangi, M. and Mitronovas, W. (1973) Dynamics of the downgoing lithosphere. *Tectonophysics* **19**, 133–147.

Weissel, J.K. (1981) Magnetic lineations in marginal basins of the western Pacific. *Royal Society, London, Philosophical Transactions* **A300**, 223–245.

Chapter 10

Elder, J. (1976) *The bowels of the Earth*. Oxford University Press.

Forsyth, D. and Uyeda, S. (1975) On the relative importance of the driving forces of plate motion. *Royal Astronomical Society, Geophysical Journal* **43**, 163–200.

McKenzie, D.P. (1969) Speculations on the consequences and causes of plate motions. *Royal Astronomical Society, Geophysical Journal* **18**, 1–32

Morgan, W.J. (1971) Convection plumes in the lower mantle. *Nature* **230**, 5288, 42–43.

— (1972) Deep mantle convection plumes and plate motions. *American Association of Petroleum Geologists, Bulletin* **56** (2), 203–213.

Ringwood, A.E. (1982) Phase transformations and differentiation in subducted lithosphere: implications for mantle dynamics and crustal evolution. *Journal of Geology* **90** (6), 611–643.

Wilson, J.T. (1963a) Hypothesis of Earth's behaviour. *Nature* **198**, 925–929.

— (1963b) A possible origin of the Hawaiian islands. *Canadian Journal of Physics* **41**, 863–870.

— (1973) Mantle plumes and plate motions. *Tectonophysics* **19**, 149–164.

Chapter 11

Krumbein, W.C. and Sloss, L.L. (1951) *Stratigraphy and sedimentation*. San Francisco, Freeman.

Lyell, Charles (1838–1865) *Elements of Geology*. London, John Murray.

Payton, C.E. (ed.) (1977) *Seismic stratigraphy – applications to hydrocarbon exploration*. American Association of Petroleum Geologists, Memoir 26, Tulsa, Oklahoma.

Miall, A.D. (1997) *The geology of stratigraphic sequences*. New York, Springer.

Mitchum, R.M., Jr., Vail, P.R. and Sangree, J.B. (1977a) Seismic stratigraphy and global changes of sea level, Part 6: stratigraphic interpretation of seismic reflection patterns. In Payton, C.E. (ed.) (1977) 117–133.

Mitchum, R.M., Jr., Vail, P.R. and Thompson, S., III (1977b) Seismic stratigraphy and global changes of sea level, Part 2: the depositional sequence as a basic unit for stratigraphic analysis. In Payton, C.E. (ed.) (1977) 53–62.

Sloss, L.L. (1963) Sequences in the cratonic interior of North America. *Geological Society of America, Bulletin* **74**, 93–113.

Sloss, L.L. (1991) The tectonic factor in sea level changes: a countervailing view. *Journal of Geophysical Research* **96B**, 6609–6617.

Sloss, L.L., Krumbein, W.C. and Dapples, E.C. (1949) Integrated facies analysis. In Longwell, C.R. (ed.) *Sedimentary facies in geologic history*. Geological Society of America, Memoir **39**, 91–124.

Vail, P.R., Mitchum, R.M., Jr., and Thompson, S., III (1977a) Seismic stratigraphy and global changes of sea level, Part 3: relative changes of sea level from coastal onlap. In Payton, C.E. (ed.) (1977) 63–81.

Vail, P.R., Mitchum, R.M., Jr., and Thompson, S., III (1977b) Seismic stratigraphy and global changes of sea level, Part 4: global cycles of relative changes of sea level. In Payton, C.E. (ed.) (1977) 81–97.

Wheeler, H.E. (1958) Time stratigraphy. *American Association of Petroleum Geology*, Bulletin **42**, 1047–1063.

Chapter 12

Beloussov, V.V. (1962) *Basic problems in geotectonics*. New York, McGraw Hill.

Bott, M.H.P. and Kusznir, N.J. (1981) The origin of tectonic stress in the lithosphere.

Tectonophysics **105**, 1–13.

Dahlmayrac, B. and Molnar, P. (1981) Parallel thrust and normal faulting in Peru and constraints on the state of stress. *Earth and Planetary Science Letters* **55**, 473–481.

Dewey, J.F. (1988) Extensional collapse of orogens. *Tectonophysics* **7**, 1123–1139.

Houseman, G.A., McKenzie, D.P. and Molnar, P. (1981) Convective instability of a thickened boundary layer and its relevance for the thermal evolution of continental convergence belts. *Journal of Geophysical Research* **86**, 6115–6135.

Law, R.D., Searle, M.P. and Godin, L. (eds) (2006) *Channel flow, ductile extrusion and exhumation in continental collision zones.* Geological Society, London, Special Publications **268**.

McKenzie, D.P. (1978) Some remarks on the development of sedimentary basins. *Earth and Planetary Science Letters* **40**, 25–32.

Ramberg, H. (1963) Experimental study of gravity tectonics by means of centrifuged models. *Geological Institutions of the University of Uppsala, Bulletin* **42**.

— (1967) *Gravity, deformation and the Earth's crust.* Academic Press, London.

Royden, L.H. and Burchfiel, B.C. (1987) Thin-skinned extension within the convergent Himalayan region: gravitational collapse of a Miocene topographic front. In Coward, M.P., Dewey, J.F. and Hancock, P.L. *Continental extensional tectonics.* Geological Society of London, Special Papers **28**, 611–619.

Trusheim, E. (1960) Mechanism of salt migration in Germany. *American Association of Petroleum Geologists, Bulletin* **44**, 1519–1540.

Umbgrove, J.H.F. (1950) *Symphony of the Earth.* The Hague, Martinus Nijhoff.

Van Bemmelen, R.W. (1954) *Mountain building.* Dordrecht, Springer.

Wernicke, B. (1981) Low-angle normal faults in the Basin and Range Province: nappe tectonics in an extending orogen. *Nature* **291**, 645–648.

Zeck, H.P. (1999) Alpine plate kinematics in the western Mediterranean: a westward-directed subduction regime followed by slab roll-back and slab detachment. In Durand, B., Jolivet, L., Horvath, F. and Séranne, M. (eds) *The Mediterranean basins: Tertiary extension within the Alpine orogen.* Geological Society of London, Special Publications **156**, 109–120.

Index

Page numbers in *italic* denote figures.

Index

Index

Index